Cloridorme Formation, Middle Ordovician Flysch, Northern Gaspé Peninsula, Quebec

Paul Enos

Shell Development Company, Exploration and Production Research Division, Houston, Texas

THE GEOLOGICAL SOCIETY OF AMERICA

SPECIAL PAPER 117

Library of Congress Catalog Card Number 76-81692

S.B.N. 8137-2117-2

Published by

THE GEOLOGICAL SOCIETY OF AMERICA, INC.

Colorado Building, P. O. Box 1719

Boulder, Colorado 80302

Printed in The United States of America

The printing of this volume has been made possible

through the bequest of

Richard Alexander Fullerton Penrose, Jr.,

and is partially supported by a grant from

The National Science Foundation.

Acknowledgments

Dr. John E. Sanders suggested the exposures of the northern Gaspé coast for study of flysch sedimentation, gave advice on conducting the work, and made many helpful criticisms on the dissertation on which this paper is based.[1] Dr. Sanders and Dr. John Rodgers visited the area during the field work. Erno Bonebakker and William Schultz assisted with the field work in 1962 and 1963, respectively, on the National Science Foundation Undergraduate Research Participation Program. Graptolite identifications were made by Dr. William B. N. Berry. Technical assistance of Malcolm McConnell, Wallace Phelps, and especially Mrs. Mary Ann Doolittle was indispensable. Critical reading of the manuscript by Dr. Sanders and by Dr. Earle F. McBride is greatly appreciated. The most valuable assistant in this study was my wife, Carol, who journeyed to Gaspé each summer, typed the preliminary dissertation manuscripts, and provided encouragement throughout the work.

Financial assistance was provided by the National Science Foundation in form of Graduate Fellowships in 1961 and 1962 and a Summer Fellowship in 1963, by Sigma Xi-RESA Grants-in-Aid in the summers of 1961 and 1963, and by a Penrose Grant from the Geological Society of America in the summer of 1962.

[1]This paper represents part of a dissertation submitted August 1964, to Yale University, New Haven, Connecticut, in partial fulfillment of the requirements for the degree of Doctor of Philosophy.

Contents

FIGURES

PLATES

TABLES

Abstract

Cloridorme Formation is proposed as a new name to replace the time-stratigraphic usage of Normanskill Formation for the upper Middle Ordovician rocks of northern Gaspé Peninsula. Detailed physical stratigraphy in excellent exposures along the Gaspé coast permits subdivision of the formation, clarifies the structural pattern of the coastal belt, and delineates the source area of these rocks.

Graywacke, calcisiltite, calcareous wacke, and some carbonate rocks form thin interbeds in dark-gray argillite with an exposed thickness of about 7700 m. Valley-and-ridge folds were formed in the Taconic orogeny (Late Ordovician), accompanied by the formation of three structural blocks bounded by major faults. The stratigraphic sequence is recognizable within each block, but relations between blocks are uncertain. With a few exceptions the lithologies of the Cloridorme Formation recur in several units, but alternation of terrigenous clastic wedges with clastic carbonate-rich wedges permits the definition of 14 members.

Abundant sedimentary structures and reworked benthonic fossils among the pelagic autochthonous fauna indicate turbidity-current deposition of the graywacke, calcisiltite, and calcareous wacke. Current directions were longitudinal from the east except in the upper two members where they were reversed. The present limited extent and *en echelon* configuration of the members suggest that they are wedges built successively toward the west as the elongate trough was filled from the eastern end.

Fragments of serpentine, volcanic rocks, chert, argillaceous terrigenous rocks, and chromite indicate cannibalism of the adjacent Cambrian and Lower Ordovician eugeosynclinal rocks to the south, as the source of terrigenous material. The carbonate-rich rocks were supplied from a carbonate shelf whose exact location is unknown. It is inferred that the carbonate source was closer to the flysch basin than the terrigenous source and was only partly isolated from terrigenous material.

The Cloridorme Formation is age equivalent to the upper part of the type Normanskill Formation. Flysch of Early or Middle Ordovician age forms a narrow, nearly continuous belt from western Newfoundland to eastern Tennessee.

Introduction

Dark-gray argillite and interbedded graywacke and calcisiltite of Middle Ordovician age are exposed in a narrow belt, 23 km in maximum width, along the northern coast of Gaspé Peninsula, Quebec, between Jersey Cove and Marsoui (Cloridorme Formation, Fig. 1). The belt continues inland near the coast as far west as Ste.-Anne-des-Monts (McGerrigle, 1954b), a total arc of 175 km. The only exposed contacts of these Middle Ordovician rocks are with a Lower Ordovician (McGerrigle, 1953) and locally Upper Cambrian sequence (Jones, 1935; Kindle, 1938; Hutchinson, 1952; McGerrigle, 1954b) of similar but considerably more varied lithofacies (Cap-des-Rosiers Formation, Fig. 1). In addition to dark argillite and graywacke, the older sequence contains red and greenish-gray shale, chert, clean quartzite, limestone conglomerate, and volcanic rocks. These older rocks are exposed in a belt up to 38 km wide along the northern Gaspé coast and inland from the Middle Ordovician rocks. Rocks of the Appalachian eugeosyncline of these ages and lithofacies are preserved in a narrow belt in the Appalachian fold belt from northwestern Newfoundland to southeastern Tennessee (Fig. 2).

These rocks are of the type commonly called flysch. Flysch is used here to indicate a thick sequence (thousands of feet or several hundreds of meters) of alternating thin beds of pelagic shale and coarser clastic rocks. Shales of flysch sequences are characteristically structureless or parallel laminated and may be calcareous or silty. The coarser clastic rocks commonly are in graded beds with some combination of scour marks, tool marks, ripple cross-lamination, convoluted lamination, shale clasts, and the like, considered characteristic of turbidity-current deposition.

This lithofacies definition would include the rocks of the more extended and genetic definitions of Tercier (1947) and Sujkowski (1957); it is similar to but less restrictive than the detailed lithologic characteristics of flysch described by Dzulynski (1963). It will be seen, however, that the Middle Ordovician rocks of northern Gaspé Peninsula belong to the flysch facies of any of several usages.

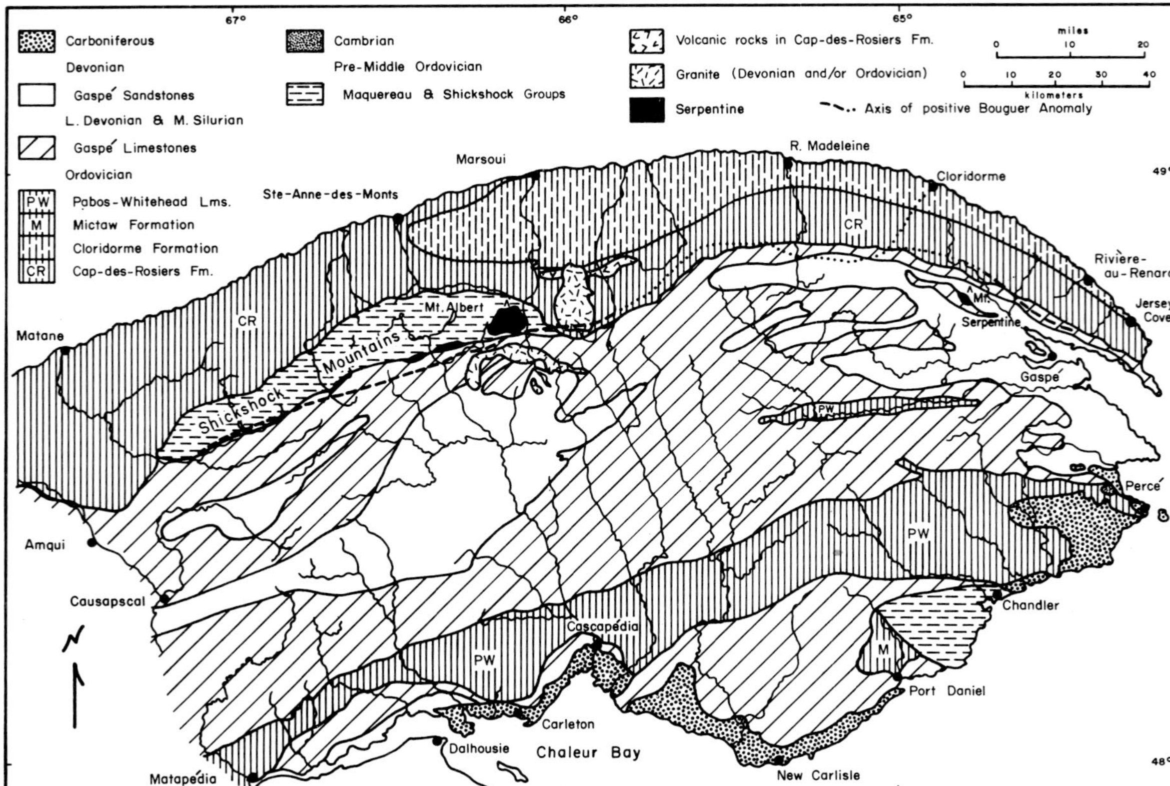

Figure 1. Geology of Gaspé Peninsula. Modified after McGerrigle (1950, 1953).

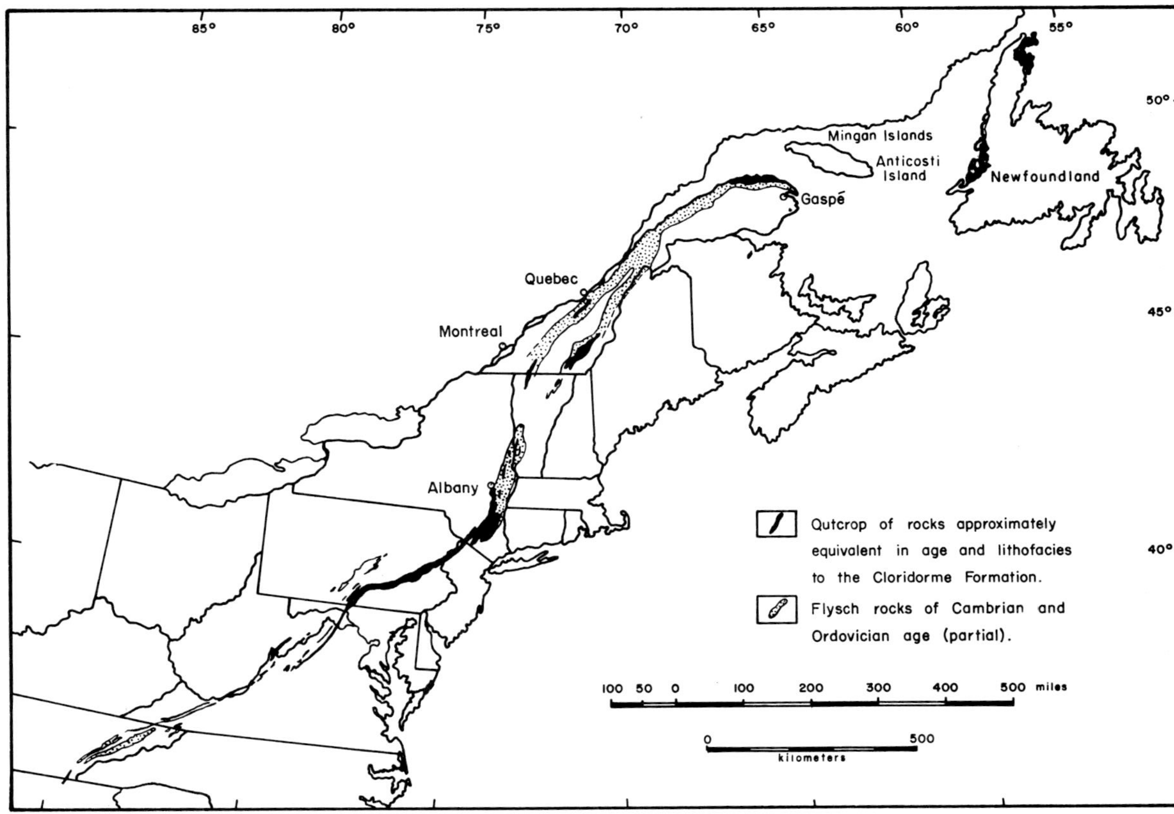

Figure 2. Cambrian and Ordovician flysch of the folded Appalachians.

OBJECTIVES AND AREA OF STUDY

The Middle Ordovician flysch was studied in the numerous coastal exposures of northern Gaspé by means of detailed and reconnaissance measured sections (Pl. 1). A tentative stratigraphy of the Middle Ordovician rocks was established, the structure of the coastal belt was mapped, and the source rocks were identified. Stratigraphic control was established from the detailed sections to the level of individual beds in some units, as a basis for analysis of longitudinal changes in turbidity-current deposits (Enos, 1965, 1969).

The coastal exposures of Middle Ordovician flysch are between the villages of Jersey Cove and Marsoui, a distance of 145 km. Reconnaissance work was done along the entire coast and in a few inland exposures. Detailed work was done between La Grande Anse (11 km west of Grande Vallée) and Pointe Route (9 km east of Grande-Vallée) and at St. Hélier (Grand Étang) 21 km to the east (Pl. 1).

METHODS OF STUDY

Nine months were spent in the field during the summers of 1961 through 1963. Of this time approximately 18 days were spent reconnoitering stratigraphy and structure, 23 days measuring reconnaissance sections, and the remaining time measuring detailed sections. A total of 3640 m of section was measured in detail, of which 1970 m are duplicate measurements of the same stratigraphic interval at different localities.

In reconnaissance sections, thickness of beds, lithologic type, maximum grain size, and orientation of current indicators were recorded in stratigraphic sequence. Any special features of individual beds were noted. Detailed sections were plotted graphically to show internal structures and weathering profile (as in Moore and others, 1951) on a vertical scale of approximately 1:40 (1 m per inch on 10x10 graph paper). Textures, structures, paleocurrent directions, and the nature of bedding surfaces were recorded for each bed using symbols and codes adapted from Bouma (1961, 1962) and Sanders (1959). The maximum grain size in phi units at the base and top of each bed was estimated. In poorly sorted rocks a realistic estimate of mean grain size is difficult, whereas the maximum can be readily estimated. The utility of the parameter maximum grain size in graywackes has been championed by Middleton (1962, p. 737) and Passega (1957, p. 1956). The type of lithology was recorded by code number. Sketches or notes on color, unusual structures, duplicate paleocurrent readings, samples, and photographs were plotted in stratigraphic position.

Orientations of current indicators were corrected for dip in the field by rotation about the strike. This assumes concentric (flexure) folding without significant plunges. Corrections for plunge necessary at a few localities were made by the graphical solution of Ramsay (1961). Partial shear folding with compression of about 10 percent was encountered near a few fold axes, but in these cases measurements could be made where dips were low and current directions were nearly parallel to axial trends of the folds so that corrections were negligible.

Field methods for removing the effects of dip in paleocurrent measurements have been described by Kuenen and Sanders (1956), ten Haaf (1959), and Bouma (1961). A dip compensator modified from Bouma's design was constructed from a standard layout protractor attached to a straight-edged plexiglass plate in which a level was mounted. The dip compensator was used to read directly the angle between strike (horizontal line in a bedding surface) and the directional feature. This angle was added to or subtracted from the strike as required to give the prefolding orientation of the feature. Correcting for dip in the field takes no more time than measuring the attitude of the feature for later correction on a stereographic net, permits a continual survey of the corrected directions, and eliminates one step, thus one potential source of error.

Separation of heavy minerals from the well-indurated sandstones proved difficult but yielded important information. A single control sample was used for eighteen different combinations of crushing, grinding, washing, elutriating, acid treating, screening, ultrasonic vibrating, and centrifuging. A separation procedure similar to that described by Hutton (1950, p. 639) was selected to give a clean separation with little grain breakage, iron contamination, or staining.

The sandstone classification of Gilbert (1954) is used because, in addition to its general utility, the position of matrix and of polycrystalline quartz in the classification appears more meaningful than in other popular classifications. The use of graywacke (as a special type of wacke) is left somewhat subjective, which seems to be a very appropriate disposition of this name at present, although it is hardly a satisfactory final solution.

REGIONAL GEOLOGIC SETTING

The Ordovician and Upper Cambrian flysch of northern Gaspé Peninsula is divided into two formations, the Cap-des-Rosiers Formation of Early Ordovician and locally Late Cambrian age (Kindle, 1938, p. 12; called "Deepkill Formation" by McGerrigle, 1953) and the "Normanskill Formation" (McGerrigle, 1953) of Middle Ordovician age, here called Cloridorme

Formation. Logan's line, the structural front of the Appalachians from New York state to Newfoundland, lies north of the flysch terrane beneath the St. Lawrence estuary. The Ordovician rocks north of the St. Lawrence River are thin, flat-lying limestones and shales resting on the Precambrian rocks of the Canadian Shield. Upper Ordovician rocks crop out on Anticosti Island (Twenhofel, 1928) and Lower to lower Middle Ordovician rocks on the Mingan Islands to the north (Fig. 2; Twenhofel, 1938). Any temporal equivalents of the Middle Ordovician flysch must lie between them, beneath the Jacques Cartier Passage. Blocks of black shale bearing Middle or early Late Ordovician graptolites found on the north coast of Anticosti Island following storms (Makasty Shale, Twenhofel, 1928) may represent such deposits.

The flysch terrane is locally bounded on the south by the crystalline massif of the Shickshock Group (Fig. 1). These schists and volcanic rocks may be metamorphosed equivalents of the Lower Ordovician and Upper Cambrian flysch (Ollerenshaw, 1961; Béland, 1957) or much older (McGerrigle, 1954b; Ells, 1884). The contact aureole of the Mt. Albert serpentine, a stock which intruded the Shickshock Group, has been dated at 495 m.y., or earliest Ordovician, by K/Ar methods (Neale and others, 1961).

The Middle Ordovician and older rocks were tightly folded in the Taconic orogeny (Crickmay, 1932; Neale and others, 1961) and later overlain with angular unconformity by Middle Silurian and Lower Devonian miogeosynclinal limestones and shales of the central Gaspé synclinorium (Kindle, 1938; Cumming, 1959). Limestone deposition was succeeded by deposition of terrigenous clastics beginning in the east in the Early Devonian and spreading throughout the area by Middle Devonian. These terrigenous clastics, collectively the "Gaspé sandstones" (McGerrigle, 1950; Logan, 1863), grade upward from drab marine deposits into continental red beds. Open folds were formed, presumably in the Acadian orogeny, before deposition during the Carboniferous of coarse continental red beds in peripheral basins underlying the present Gulf of St. Lawrence and Chaleur Bay (Alcock, 1935).

In southeastern Gaspé Peninsula the Maquereau Group of crystalline rocks, which resembles the Shickshock Group in lithology and degree of metamorphism, is overlain with a basal conglomerate by less metamorphosed slates and graywackes of the Mictaw Formation (Alcock, 1935; Ayrton, 1961). The Mictaw Formation is Middle Ordovician but contains no graptolite species in common with the Middle Ordovician rocks of similar lithofacies on the northern coast. The Mictaw rocks are overlain with angular unconformity by Middle Silurian limestone, shale, and sandstone.

PREVIOUS WORK

Geologic investigations of northern Gaspé Peninsula were begun by Sir William Logan in 1844 (Logan, 1846, 1863). He described the Ordovician flysch and noted recumbent folds in the flysch belt. The structural discontinuity now known as Logan's line was placed south of the area of "Hudson River and Utica" rocks (the Middle Ordovician flysch), separating the Middle Ordovician from the Lower Ordovician and older rocks, rather than beneath the St. Lawrence.

A geologic map of Gaspé (scale 1:253,440) was published by Ells (1883, 1884). Only recently have more detailed maps of the northern coast become available through the work of McGerrigle (1953, 1954b, 1959). The map of the Madeleine River area on a scale of 1:63,360 (McGerrigle, 1959) includes part of the area studied in detail for this report. McGerrigle briefly described the Middle Ordovician rocks but did not attempt to subdivide them.

Structure

The structural configuration of the Middle Ordovician flysch is typical of the Appalachian Valley and Ridge province; it consists of asymmetrical folds of fairly large amplitude and some thrust faults. Metamorphic grade is low but slaty cleavage is well developed in the argillaceous rocks. The Middle Ordovician rocks were broken into three structural blocks which will be referred to as the eastern, central, and western blocks. The stratigraphic sequence within each block was determined, but correlations between blocks were not possible, probably because of missing stratigraphic intervals caused by the structural dislocations separating the blocks. The boundary between the eastern and central blocks is a thrust fault that extends west from Pointe-Jaune; along it rocks of the eastern block have overridden those of the central block (Pl. 1). A transverse fault with unknown sense of displacement separates the central block from the western block at Le Grand-Ruisseau, 5 km west of Grande-Vallée.

The structure and stratigraphy were mapped in greatest detail on the central block (Pl. 1) and are at least known on the eastern block where only reconnaissance work was done. Particular care was taken to determine the plunge of folds that may have seriously altered the original orientation of current-direction structures (ten Haaf, 1959, p. 73; Ramsay, 1961). Plunge was measured directly from major or minor fold axes or was reconstructed from the stratigraphy where possible; elsewhere it was approximated by plunge of cleavage-bedding intersection.

METAMORPHISM

Axial-plane slaty cleavage is well developed in most argillaceous rocks of the Middle Ordovician Flysch. At the axis of the open syncline at Cloridorme both calcisiltite and argillite beds were thickened about 10 percent relative to the same rocks measured about 50 m from the fold axis, but such thickening is unusual. Extensive development of chlorite and sericite, on cleavage surfaces and as alteration products of feldspar, suggests metamorphism in the lower greenschist facies. Pumpellyite, present in trace amounts, was the only characteristic zeolite-facies mineral recognized.

EASTERN BLOCK

The eastern block extends along the coast from Jersey Cove to Pointe-Jaune and much further to the west inland. The only major fold mapped within this block is a steep-limbed open syncline at Anse-au-Griffon which plunges 5° east and apparently has a wavelength in excess of 3 km. Minor folds were observed in rocks superjacent to a thrust plane at St.-Hélier, along Route 6 at L'Anse-à-Valleau, near the contact with the Cap-des-Rosiers Formation at Jersey Cove, and at Rivière-au-Renard-Ouest (Pl. 2). Elsewhere the coastal exposures are of a southwest-dipping homocline.

West of Pointe-Jaune a thrust fault has carried rocks of the eastern block over the central block (Pl. 1). This fault probably extends west as far as Rivière-du-Petit-Cloridorme. A thrust fault of the same configuration, with the middle member of the eastern block overriding the lowest member of the central block, was observed at Rivière-au-Renard-Ouest (Cap aux Goélands). This may be the same thrust which strikes into the sea at Pointe-Jaune, but, if so, either the Rivière-au-Renard-Ouest exposure is a window or right lateral offset of about 2.1 km has occurred between the two exposures. A small horse of red shale occurs in the fault zone at Rivière-au-Renard-Ouest. Shale of this type is not definitely known from the Middle Ordovician flysch but occurs in the Cap-des-Rosiers Formation east of Jersey Cove.

The contact between the Middle Ordovician rocks and the Cap-des-Rosiers Formation (Lower Ordovician and Upper Cambrian) east of Jersey Cove was interpreted as a normal contact by McGerrigle (1950, 1953), but a fault contact appears more probable. Numerous southwest-dipping, low-angle thrusts of unknown displacement produce thin slices of several different lithologies at the contact. Only the lithologies of the lower two thrust slices, one predominantly of argillite and one of graywacke, resemble those found elsewhere in the Middle Ordovician terrane. Thrust slices of red shale, of green argillite, and of thinly interlaminated tan siltstone and red shale probably belong to the Cap-des-Rosiers Formation. The older Cap-des-Rosiers Formation is isoclinally folded with axial planes dipping southwest at Cap-des-Rosiers and Jersey Cove. This pattern of folding is inconsistent with a normal contact with younger rocks to the north unless successive folds are depressed toward the north as they are farther west (cross sections, Pl. 1).

Figure 1. Graywacke (type 1) and argillite of $\sigma 2$ member at Cap du Gros-Mâle near Gros-Morne, Section XX. All beds are overturned. Axial planes of the small folds at the top of the exposure dip northward, an unusual orientation. No large dislocation separates the folded beds from those beneath.

Figure 2. East-plunging minor folds at Northwest Point, Riviére-au-Renard-Ouest, Section XXI. Breached bed at observer's feet indicates plunge, about 10°. Note crumpling of beds at the fold axis. Rocks are type 1 calcisiltite and argillite of $\alpha 2$ member.

Figure 3. Type 1 graywacke (A) and parallel laminated type 2 calcisiltite (B) of $\beta 7$ member. Section II_3, looking east toward Anse à Colin, Grande-Vallée.

OUTCROPS OF CLORIDORME FLYSCH, NORTHERN GASPÉ PENINSULA, QUEBEC

ENOS, PLATE 2

CENTRAL BLOCK

The central block contains three large folds, overturned to the north, and two smaller, open folds (cross section, Pl. 1). Plunges on these folds are less than 4° except for one of 8° at St.-Yvon (4 km east of Cloridorme). Minor folds plunge up to 45° both east and west in the exposures extending about 900 m west from Pointe de Cloridorme. These may be associated with a large right-lateral fault inferred to pass near this locality. The large overturned anticline and syncline at the northern edge of the exposed part of the central block plunge 3.5°W. and 2°W., respectively, as shown by detailed stratigraphic control. No corrections are necessary on the numerous current directions measured on the mutual overturned limb of these folds because of the small plunge (Ramsay, 1961).

Two sets of transverse faults with predominantly strike-slip movement occur in the central block. Faults of a right-lateral set strike about N.20°W.; those of a left-lateral set about N.60°E. The orientation and sense of movement on the right-lateral faults suggest one set of a very unequally developed conjugate pair forming an acute angle to the maximum stress direction inferred from the folds. However, the left-lateral set does not fit this pattern, and both sets appear to have formed at about the same time because there is no consistent age relation between them. The sense of the last movement on several faults was determined from slickensides that plunge about 20°N. In most of the faults of both sets the horizontal component of movement is a few tens of meters or less, but larger ones, all right lateral, were mapped at Anse à Colin, Grande-Vallée (50 m horizontal component); at Anse à Dery, Petite-Vallée (140 m); west of Pointe de Cloridorme (inferred 350 to 900 m); and at Le Grand-Ruisseau 5 km west of Grande-Vallée (inferred 1700 m). The fault at Le Grand-Ruisseau forms the boundary between central and western blocks.

WESTERN BLOCK

The largest fold in the western block, and the one extending farthest east toward the central block, is an anticline which plunges 2°W. at Manche-d'Épée (22 km west of Grand Vallée) and 8°E. at Madeleine-Centre. The north limb of the asymmetrical anticline is overturned throughout most of its exposed length, but at Manche-d'Epée it is upright and dips 40°N. The smaller folds to the north and west of this anticline have one broad, gently south-dipping limb and a short, nearly vertical or slightly overturned limb so that the traces of anticlinal and synclinal axes are nearly superposed. The folds thus resemble a flight of stairs descending toward the north (cross sections, Pl. 1). Axial plunges as great as 20°W. in folds of this type at Mont-Louis necessitated corrections in current-direction readings in section XXIII.

From Mont-Louis west to Marsoui flexures trending north-south are superposed on the dominant west-southwest structural trend. Shallow basins and domes are the common expression of the north-south trend, but sharp, north-plunging minor folds exist between Rivière-à-Claude and Marsoui.

Several high-angle longitudinal faults, apparently of small displacement downthrown to the north, occur along the contact between γ_2 and γ_3 members seen on both sides of the cove at Anse-Pleureuse (6 km east of Mont-Louis).

McGerrigle (1953) mapped a normal contact at Rivière-à-Claude with Middle Ordovician ("Normanskill") rocks on the south and Cap-des-Rosiers ("Deepkill") rocks on the north. An apparently conformable homoclinal sequence, upright and dipping 40°NW. occurs here, so the Cap-des-Rosiers Formation could not lie north of the younger Middle Ordovician rocks in normal sequence. The general homoclinal trend continues from southeast of Rivière-à-Claude westward to Marsoui except where interrupted by gently north-plunging cross folds. The probable contact at Marsoui is covered, but fine-grained, buff-colored clastic dolomite and highly sheared green mudstone of the Cap-des-Rosiers Formation crop out west of Marsoui, although an outlier of Middle Ordovician rocks is shown here by McGerrigle (1953). The rocks west of Rivière-à-Claude are interpreted to be the youngest exposed Middle Ordovician rocks, so either the stratigraphic sequence as presented below is grossly wrong at this point or the contact at Marsoui represents a major structural dislocation. It is clear at least that these rocks are Middle Ordovician in age and not part of the Cap-des-Rosiers Formation; Trenton graptolites were collected 2.7 km east of Marsoui (p. 40; Trenton of Cooper, 1956; identification by W. B. N. Berry). The fault which brings these presumably youngest Middle Ordovician rocks against the Cap-des-Rosiers Formation at Marsoui may be a transverse fault which extends up Rivière Marsoui. The Middle Ordovician rocks and the Cap-des-Rosiers Formation are intimately infolded west of Marsoui as mapped by McGerrigle (1954b); thus it is unlikely that a major fault extends west along their contact.

Stratigraphy

The Middle Ordovician flysch of northern Gaspé Peninsula, previously assigned to the Normanskill Formation (McGerrigle, 1953, 1954b, 1959), is here named the Cloridorme Formation. The name Normanskill was extended to Gaspé Peninsula largely on a time-stratigraphic basis from a poorly defined unit with a type locality near Albany, New York (Ruedemann, 1901; Berry, 1962b). Some lithologic similarities exist between the New York and Gaspé rocks, but in the intervening 800 miles rocks of the particular age and lithology of the Normanskill are rare. Only a few exposures of flysch as young as Middle Ordovician are recognized between Ste.-Anne-des-Monts on Gaspè Peninsula and Quebec City, a distance of 300 miles (Claude Hubert, 1966, personal commun.) Correlations southward along the Green Mountain-Sutton Mountain axis are complicated by areas of complex structure and extensive metamorphism. The time-stratigraphic use of the term Normanskill on Gaspé Peninsula and the apparent lack of continuity with the New York sequence makes the continued use of this name for the Gaspé rocks undesirable.

The name Cloridorme Formation is taken from the villages of Cloridorme and Cloridorme-Ouest on Baie de Cloridorme at the mouths of the Petit and Grand Cloridorme Rivers, Cloridorme Township, Quebec. The type locality is along the St. Lawrence estuary between Cap-des-Rosiers and Marsoui. The Cloridorme Formation is subdivided informally into fourteen members with more specific type localities. The members fall into three larger "sequences." Each sequence is entirely confined to one of the three structural blocks; thus the relative stratigraphic positions of the sequences and the thickness of intervals faulted out between blocks is unknown.

Normal contacts of the Cloridorme Formation with the underlying Cap-des-Rosiers Formation were not observed in this study, but they have been mapped in the Tourelle area, southeast of Ste.-Anne-des-Monts (McGerrigle, 1954b). No younger rocks are found in contact with the Cloridorme Formation. The exposed part of the formation is estimated to be 7700 m thick, although this figure may be in error by as much as 100 percent.

LITHOLOGIES

The Cloridorme Formation consists of 60 percent dark, slaty argillite. Interbedded are coarse graywackes (15 percent), calcereous wackes (3 percent), and calcisiltites (20 percent), all of which are attributed to turbidity-current deposition. Dolostone, limestone, volcanic ash, and silty dolomitic argillite of pelagic origin total about 2 percent of the formation. Members are defined primarily by relative abundance and sequence of lithologies rather than by unique occurrences. It is therefore convenient to first describe the lithologies and to then use them as building blocks in defining the members. The lithologies are discussed in the following order: argillite, calcisiltite (two types), calcareous wacke, graywacke (three types), and the four minor lithologies. Compositions of the lithologic types are summarized in Table 1 and the sedimentary structures in Tables 2 and 3.

Argillite

Argillite is used here for erstwhile shale with coarse clots of mica and with slaty cleavage which obscures bedding. Petrography of the argillite was limited to examination of outcrops and of thin sections of eroded clasts redeposited in graywacke beds because the slaty cleavage makes thin-section preparation difficult.

Chlorite and illite, identified with the microscope, comprise most of the argillite. Angular quartz grains, 0.005 to 0.02 mm in diameter, are rare. Pyrite nodules and isolated pyrite euhedra are common. Disseminated murky organic matter and fine-grained pyrite apparently account for the dark-gray shade of the argillite. Calcareous, dolomitic, and silty argillite are rare but occur as vertical or lateral transitions to limestone, dolostone, and graywacke, respectively. Flakes of chlorite and muscovite as large as 0.5 mm lie in the cleavage planes. A few graptolites were found compressed in bedding planes, and they would probably be found in abundance as they are in the graywacke if the rock split along bedding rather than along cleavage. Nautiloid cephalopods were the only other fossils found in the argillite.

Calcisiltite

Clastic carbonates containing various amounts of quartz are present in all fourteen members, and they constitute most of the allochthonous (nonpelagic) portion of some members. Two types of calcisiltite are distinguished on the basis of internal structure, bedding thickness, and clay-matrix content. The more abundant type 1 calcisiltites are characterized by thin beds (Pl. 2), by ripple-drift cross-lamination and convoluted lamination, and by relatively little matrix. Type 2 calcisiltites form thick beds with climbing ripple-drift cross-lamination that produces false bedding.

Beds of type 1 calcisiltite appear continuous in any one outcrop but cannot be satisfactorily traced between measured sections because they lack distinguishing features. The bases of many beds are dotted by myriad small hemispheric protrusions of sand 2-5 mm in diameter which reflect depressions, apparently burrows, in the mud surfaces on which they were deposited.

Ripple-drift cross-lamination (Sorby, 1859), typically 0.5 to 2 cm in amplitude, is found in 2 to 50 percent of the type 1 calcisiltite beds, varying with the unit and with the degree of exposure. This is commonly climbing ripple-drift cross-lamination as described by Walker (1963), that is, the ripples climb progressively onto the stoss slope of the ripple immediately downstream; but the angle of climb varies considerably, depending upon the amount of fallout from suspension which controls the erosion of the stoss slopes (Sanders, 1963). Truncated fore-set lamination oversteepened by current drag (Wood and Smith, 1958; Sanders, 1960) is common and grades into convoluted lamination in as many as 20 percent of the cross-laminated beds. A sequential development from cross-lamination at the base of a bed through oversteepened cross-lamination, convoluted lamination, oversteepened cross-lamination, and back to cross-lamination at the top is seen in many beds. Convoluted lamination also appears to have developed from current drag on planar lamination, suggesting that some planar lamination may be formed in the flow regimes above the first critical velocity (Gilbert, 1914, p. 33) if the analysis of traction currents is applicable in the deposition of these calcisiltites. However, it is difficult to establish the original form of most convoluted lamination unless a transition is exposed.

Planar lamination or complete lack of lamination in type 1 calcisiltites is nearly confined to strata less than 2 cm thick within predominately graywacke members. Internal lamination is defined by thin concentrations of chlorite or carbonaceous material to give a dark layer best seen on weathered surfaces. Despite this inhomogeneity, few such laminations weather out as argillaceous partings or zones of splitting. No graded beds of type 1 calcisiltite were detected in the field or in thin sections. A few beds are darker at the top, suggesting an increase in argillaceous content, hence grading, but most are uniform in color and in distinctness of contact at the top and base.

Type 1 calcisiltite is well indurated and breaks with a smooth, subconchoidal fracture, but the weathering profile is controlled by vertical joints which produce flat faces with angular corners. Most of the terrigenous grains are 0.01 to 0.09 mm in diameter. The carbonate grains are largely irregular single crystals, 0.005 to 0.08 mm, but in every specimen a few percent of the grains are polycrystalline, composed of many irregular particles less than

0.004 mm in diameter (micritic, Folk, 1959). These suggest that the rock has undergone considerable recrystallization, forming single crystals from polycrystalline carbonate mud particles of about the same size. The terrigenous grains, largely quartz, are angular, equant to slightly elongate, and moderately well sorted. Many of them are embayed or partially replaced by carbonate. The carbonate grains are less uniform in size, are irregular in outline, and enclose quartz grains in a few cases, probably all as a result of recrystallization. The carbonate content of type 1 calcisiltites varies between 45 and 82 percent (Table 1). A sample analyzed by X-ray diffraction contained 33 percent dolomite and 47 percent calcite, but some rocks do not give a noticeable reaction with cold HC1, which suggests a much higher proportion of dolomite to calcite. Only a few dolomite rhombs can be distinguished in thin section. The matrix, which varies between 2 and 32 percent, is largely chlorite with some carbonaceous material. The heavy minerals include pyrite, leucoxene, garnet, zircon, and chromite. The only fossils found in these rocks were a few graptolites, gastropods, inarticulate brachiopods (?), and nautiloids.

Calcisiltites of type 2 lithology are distinguished from type 1 calcisiltites by the presence of parallel lamination or false bedding, or both, by an average matrix content of 18.0 percent versus 11.6 percent in type 1 calcisiltites, and by an average thickness of 24 cm, about seven times that of the type 1 calcisiltites. Despite their considerable thickness, type 2 calcisiltite beds are less persistent laterally than any other lithology of the Chloridorme Formation; none could be traced as far as 1 mile.

Some type 2 calcisiltites form the top of graded graywacke beds, in which case the base may be fairly distinct or gradational from fine-grained carbonate-rich graywacke. Either parallel or cross-lamination (Pl. 2) is present in all type 2 calcisiltites and generally forms distinct partings or splittings (McKee and Weir, 1953). Parallel lamination is formed by brown calcisiltite laminae less than 0.25 cm thick alternating with gray silty argillite laminae 0.25 to 1 cm thick. In some cases argillite laminae become relatively thicker toward the top of a bed, perhaps as a subtle form of graded bed.

Ripple-drift cross-lamination was formed with abundant fallout of grains from suspension so that each ripple climbed up the stoss slope of the next ripple downcurrent with little or no erosion of the back-set laminae (Walker, 1963). Fine-grained micaceous material accumulated on the lee faces of the ripple—or on the crests or in the troughs in some cases—to produce a uniform layer of dark argillite which dips upcurrent at the angle of climb, generally about 15°. The effect in an outcrop in which the true lamination is obscure is a striking semblance of large-scale planar cross-bedding dipping in exactly the opposite direction to the actual direction of current flow (Fig. 3). This apparent stratification produced by the regular alternation of finer grains in the fore-set beds and coarser grains in the back-set beds is a form of false bedding. The accumulation of fines in the fore-set beds is produced by winnowing action on the stoss slopes where the current velocity is greatest and by accumulation of the finer material that is swept over the crests into the current shadow on the lee faces (Menard, 1950). Rapid fallout is essential in this analysis, for the ripples must advance primarily by addition of new material; if all grain sizes are reworked even the length of one ripple, the coarser ones could accumulate preferentially in the troughs as in the false bedding of Spurr (1894). The common accumulation of fines on the face slopes instead of in the ripple troughs may be explained by flow separation in the troughs with formation of backward-directed eddies (Jopling, 1964) which carry the fines partially up the face slopes (Allen and Narayan, 1964).

The median grain size of both the terrigenous and the carbonate grains is about 0.02 mm. The quartz grains are moderately well sorted, but locally very fine-grained quartz is present. The carbonate grains are dominantly single crystals which may enclose quartz grains, but a few of them are polycrystalline fragments consisting of many grains less than 0.004 mm in diameter. This suggests that the larger single crystals were recrystallized from microcrystalline aggregates of about the same size and that a few of them grew to enclose other grains.

Calcareous Wacke

Calcareous wackes contain greater than 10 percent argillaceous matrix (wackes of Gilbert, 1954) but contain more sparry calcite cement than matrix. Rocks of this type are not found in most recent sandstone classifications (Klein, 1963) because argillaceous matrix and sparry carbonate cement are generally considered "antipathetic." In these rocks, however, it appears that the matrix and cement have the same significance in terms of transportation and sorting because much of the sparry cement is recrystallized lime-mud matrix. Calcareous wackes grade vertically and perhaps laterally into type 1 calcisiltites by decrease in both grain size and degree of recrystallization of carbonates.

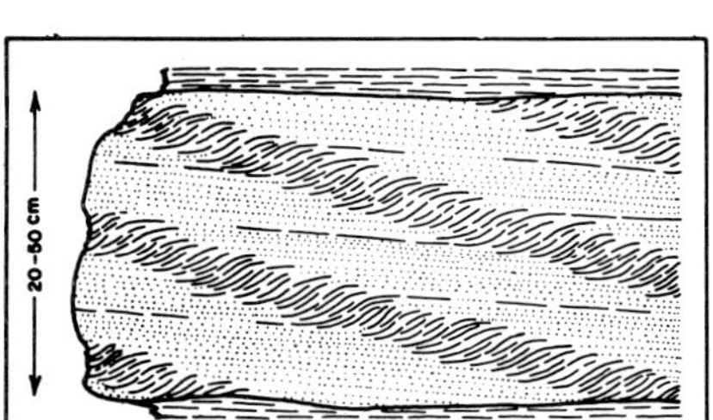

Figure 3. Climbing ripple-drift cross-lamination in type 2 calcisiltite. Current direction from right to left. With development of slaty cleavage or in poor exposure the concentration of argillite in foresets produces a semblance of planar medium-scale cross-bedding dipping to the right.

Calcareous wacke beds are very persistent laterally within and between outcrops. A characteristic sequence of lamination within beds (Fig. 4) is cross-lamination in a lenticular basal layer, 0 to 4 cm thick, which appears to represent migrating isolated ripples or a thin traction carpet; wavy lamination or convoluted lamination formed from current drag on wavy lamination; and planar lamination at the top. This sequence could be termed a c-d sequence of Bouma (1961, 1962), but the coarse grains and marked grading associated with cross-lamination in the basal portion are anomalous.

About 25 percent of the calcareous wackes are overlain by an argillaceous interval 2 to 40 cm thick. This is separated from the graded calcareous wacke by a distinct shale parting or a transition several centimeters thick consisting of argillaceous layers intercalated with the parallel-laminated calcisiltite at the top of the graded wacke bed. Those argillaceous intervals appear to form a genetic unit with the calcareous wackes; if coarse grains persist to the top of a wacke, they are also very sparsely distributed through the lower part of the argillaceous bed. Irregular or spindle-shaped light-brown calcareous masses are scattered through the gray argillaceous matrix. These may be concretionary in part, but contorted lamination which is visible in a few indicates that these are fragments eroded from unconsolidated calcisiltite beds. Argillite or calcisiltite fragments are rare in the graded part of calcareous wackes, but flutes and perhaps the argillaceous matrix indicate that erosion of the underlying sediment did occur preceding deposition of the wacke.

The cement at the base of graded beds is sparry calcite despite 3 to 21 percent argillaceous matrix. The large sparry calcite crystals enclose terrigenous grains, but as the size of the grains decreases upward so does the size of the carbonate crystals until the texture is a fine carbonate mud with local polycrystalline clots, just as in the calcisiltites. Thus it appears that the cement at the base results from recrystallization of microcrystalline carbonate which was transported with the terrigenous grains and that recrystallization was impeded by finer grain size with reduced permeability toward the top. The size of the transported carbonate grains probably varied from that of the

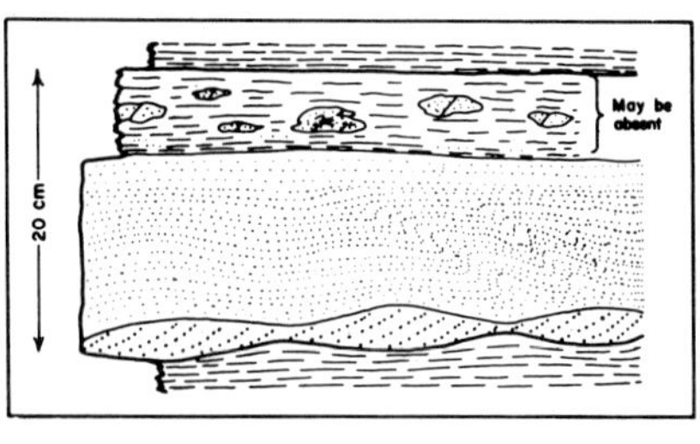

Figure 4. Sequence of sedimentary structures in a typical calcareous wacke bed. Cross-lamination at base, wavy lamination passing laterally in convoluted lamination, parallel lamination at top. Argillace top portion containing calcisiltite fragments is present in 25 percent of the beds.

coarser clastic components down to perhaps that of the fine-grained matrix, in which case the finer carbonate particles indicate textural immaturity or lack of sorting just as the argillaceous matrix does. If so, the designation of this carbonate-cemented rock as a wacke is not so paradoxical as it first appears and the graduation of calcareous wacke into calcisiltite is analogous to the graduation of terrigenous wacke into terrigenous siltstone.

Calcite has extensively replaced quartz and feldspar grains, and most feldspar grains are to some degree altered to chlorite or sericite as well. Fossils form about 2.1 percent of the coarse-grained portion of the calcareous wackes, but only a few brachiopods and echinoderms were not abraded beyond recognition.

Graywacke

Gray, well-indurated sandstones containing more than 10 percent matrix are the dominant clastic rock type in six members of the Cloridorme Formation. These are graywackes by Gilbert's (1954) classification as well as by most other recent definitions of graywacke (van Andel, 1958; Bokman, 1955; Crook, 1960; Dapples and others, 1953; Dunbar and Rodgers, 1957; Krynine, 1948; McBride, 1962; Packham, 1954; Pettijohn, 1954), including the definitive one, "It's gray and you have to wacke like hell to knock off a piece" (anonymous). They are not, however, graywackes by the classification of Folk (1954), because igneous rock fragments plus feldspar generally exceed metamorphic rock fragments, including polycrystalline quartz. Three types of graywacke were distinguished in the field on the basis of sedimentary structures, approximate matrix content, distinctness of contacts and color; these are called types 1, 2, and 3.

Type 1 graywackes, the most abundant type (Table 4), are characterized by relatively little matrix (17 percent), numerous graded beds, distinct contacts, and light color (Tables 1 to 3). The matrix content of graywackes increases downcurrent so that type 1 graywackes grade into type 2 which pass abruptly into current-deposited argillite. At the upcurrent end distinct graywacke beds may appear abruptly with local erosion or they may grade into argillite as at the downcurrent termination. Lateral variations of as much as 100 percent in thickness of beds are seen in most large outcrops.

The tops of graywacke beds may be obscured by abundant argillite fragments or by argillite lenses. Current sorting apparently concentrated the argillite fragments toward the top and current agitation may have caused loss of internal cohesion in the fragments, resulting in a muddy slurry now seen as an argillite lens. In such beds the location of the top commonly is defined by a horizon with persistent lenses of clastic grains—terrigenous or carbonate—above the argillitic interval.

TABLE 4. LITHOLOGY OF MEMBERS*

Member	Estimated thickness (m)	Total measured (m)	Argillite	Dolomitic silty argillite	Dolostone and limestone	Calcisiltite 1	Calcisiltite 2	Calcareous wacke	Graywacke 1	Graywacke 2	Graywacke 3
γ4 (est.)	510	--	35	--	--	5	--	--	60	--	--
γ3	475	63	70.2	tr.	4.3	13.4	--	9.4	2.0	0.5	0.2
γ2	800	400	41.0	--	0.9	4.3	12.1	0.1	37.5	4.1	--
γ1	550	209	26.0	57.4	3.2	3.2	--	0.8	0.3	--	9.2
β7	835	1500	44.1	--	0.7	6.7	9.6	0.6	31.1	7.3	--
β6	255	400	78.4	--	2.5	15.9	0.4	0.6	2.0	0.1	0.1
β5	80	61	46.5	--	0.6	4.3	5.4	--	35.8	7.5	--
β4	78	89	64.3	--	0.5	28.7	3.4	1.8	0.9	0.4	--
β3 (est.)	43	--	40	--	--	10	25	--	25	--	--
β2	490	246	77.4	--	0.2	17.7	tr.	4.2	0.3	--	0.2
β1	1175	561	70.4	--	1.9	7.2	--	6.7	--	0.3	13.5
α3	2000	52	51.9	11.3	1.0	18.1	--	17.7	--	--	--
α2	250	68	55.8	--	0.8	41.9	--	1.5	--	--	--
α1 (est.)	50	--	50	--	--	--	10	--	40	--	--

*Percentage of each lithology based on average measured.

Lenses or layers of calcisiltite at the tops of graywacke beds originate by disaggregation of unconsolidated clasts as suggested for the argillite lenses. Many beds contain irregular fragments of calcisiltite, with lamination distorted before lithification, scattered throughout the bed but generally increasing in abundance upward. Larger lenses or layers of calcisiltite at the tops of beds coincide significantly with the occurrence of calcisiltite fragments but have undistorted ripple-drift cross-lamination, showing a lack of cohesion during deposition. The concentration of carbonate mud to produce the lenses probably resulted from both current sorting of fine carbonate grains and current disaggregation of the unlithified calcisiltite fragments eroded locally from the bottom. Much of the argillaceous and carbonate matrix of the graywackes probably originated in the same manner.

Sixty-eight percent of type 1 graywacke beds are graded, not counting "content graded beds." In "content graded beds" (McBride, 1962, p. 50) maximum grain size does not vary from base to top, but the abundance of coarse grains decreases and the percentage of fine matrix increases so that the median grain size must decrease. Composite graded beds, with two or more graded intervals, when traced laterally can be seen to be formed by successive depositional events of which the upper one eroded any mud deposited between them. The contacts between the successive graded intervals in composite beds are very irregular and may even show "flame structures" of finer sand into the coarser overlying bed. A few beds which coalesce within an outcrop with the disappearance of the argillite between them do not form composite graded beds but appear to be homogeneous. These may be formed by lobes or pulses (Kuenen, 1953) of a single current, and the shale lens between them may also be a local turbidity-current deposit.

If a fixed sequence of internal sedimentary structures exists in these graywacke beds, it consists of only two parts, a graded basal part and a cross-laminated top of calcisiltite which is absent more commonly than present. Parallel lamination and convoluted lamination are rare. Thus Bouma's (1961, 1962) sequence of sedimentary structures is not applicable.

Poor sorting is indicated by the matrix content of 17 percent and by standard deviations of 0.85 to 1.2 phi units in the size of quartz grains alone measured in thin section. The validity of matrix as an index of sorting and a criterion for textural maturity is diminished by the origin of an undetermined amount of argillaceous interstitial material through the decomposition of unstable rock and mineral fragments (Cummins, 1962; Potter and Pryor, 1961, p. 1202). Material which is entirely clay and chlorite but which appears to have relict grain outlines, a granular texture, or ghosts of twinning is repeatedly encountered in modal analyses of these rocks. That the distinction between argillaceous matrix and fine-grained carbonate cement is dubious was discussed with the calcareous wackes.

The quartz grains of medium sand size are somewhat elongate in cross section and subangular; the sphericity index is 0.75 and roundness is 0.35, based on visual estimates (Krumbein, 1941). Some quartz grains are well rounded, although irregularities of the grain boundaries due to solution and replacement by calcite make estimates of original roundness difficult.

Graptolites are the most abundant fossils in the graywackes; they are also an important polar directional indicator and a micalike fabric element.

They are flattened in the plane of bedding to form a parting where they are concentrated because of shape and specific gravity. Despite this sorting, graptolites are not noticeably concentrated at the tops of beds and in many beds they are more abundant at the base, perhaps because a continuous supply was derived from local bottom mud by erosion. Abraded fragments of calcareous fossils are concentrated near the base of graywacke beds, apparently by high specific gravity and large size relative to detrital grains. These include, in order of decreasing abundance, brachiopods, bryozoans, nautiloids, gastropods, corals, pelecypods, crinoids, and other echinoderms.

Type 2 graywackes contain still more matrix (50 percent) than do the typical Cloridorme graywackes described above, which explains their incipient slaty cleavage and distinctive ragged weathering profile. They are also much darker because of the pyrite, metamorphic chlorite, and carbonaceous material which accompany the increase in matrix. The distinction of type 2 graywackes is useful but artificial because they are an intermediate form in the lateral gradation of type 1 graywacke into argillite, as was demonstrated by walking out numerous beds. Lateral variation in thickness of type 2 beds appears greater than in type 1 graywackes, probably due to the type 2 beds' positions near the distal end of turbidity currents and to differential compaction of the very argillaceous matrix. Distinctly graded beds are uncommon because coarse grains tend to be widely scattered throughout the muddy matrix. In many such beds, however, the abundance of coarse grains decreases slightly and the amount of matrix increases from base to top to form "content graded beds" (McBride, 1962, p. 50). Large fragments of argillite and calcisiltite, unconsolidated at the time of sedimentation, are common, especially in the upper parts of beds. Lenses of argillite, calcisiltite, and very coarse sand are distributed in the same manner and probably represent fragments which have lost much of their internal cohesion in reworking by the current. Argillaceous or carbonate mud matrix would result from continued reworking plus mixing with coarser current-borne grains.

The argillaceous matrix is composed of considerable illite with some chlorite, carbonaceous material, and very fine-grained quartz. The fine-grained carbonate, which also appears to have originated as detrital mud, is quite variable in amount. It appeared early in the field work that, as the argillaceous matrix content increased, the carbonate content decreased with the decrease in pore space, as is commonly the case in sandstones. However, many Cloridorme sandstones do not fit this pattern; muddy rocks which weather with the rough surface typical of type 2 graywackes may contain abundant carbonate. These rocks are difficult to classify as either type 1 or 2; five modal analyses are averaged separately in Table 1 as graywacke, type 2, carbonate rich. Although the matrix content (17.4 percent) is much lower than in typical type 2 graywackes, it is high relative to the terrigenous detrital grains, and if the 43.5 percent carbonate content is regarded as detrital mud, these rocks are texturally identical to other type 2 graywackes.

The mineralogies of the coarse detrital grains of the intergradational type 1 and type 2 graywackes are of course similar. Sorting by shape could explain the larger percentage of mica (4.9) in type 2 rocks. Authigenic pyrite would be favored by the increase in matrix with its carbonaceous material.

Type 3 graywackes are distinguished by very thick resistant beds, abundance of calcisiltite fragments, coarseness of grains, and sequence of internal structures. Well-indurated pebbly graywacke forms a basal layer or lenses, up to 10 cm thick, in the sequence. Dark-gray fissile argillite containing flattened argillite clasts and scattered coarse grains produces a recess in the weathering profile 2 to 8 cm thick above the resistant basal layer. Resistant, ragged-weathering, slightly dolomitic, silty argillite containing numerous large calcisiltite masses and, locally, scattered coarse grains make up most of each bed. The calcisiltite masses, as large as 25 cm across, probably were eroded locally from the bottom and were unconsolidated at the time of redeposition. Several cycles of grading in size and abundance of the fragments are seen in most beds. Highly contorted internal lamination and flattened spindle shapes are recognizable in a few fragments. The consistent shape, internal structure, orientation, and grain-size distribution of the "sandstone whirlballs" of the Carpathian Flysch (Dzulynski and others, 1957) are lacking.

Carbonate clasts make up as much as 70 percent of the grains in the coarse basal layer. The carbonate rocks include lime mudstones as the most abundant, wackestones, packstones, and a few grainstones and dolostones (classification of Dunham, 1962). It was not possible to determine petrographically whether the carbonate clasts were lithified at the time of deposition. Fossil fragments in the coarse layers include brachiopods, crinoids, other echinoderms, bryozoans, and a few graptolites. The matrix consists of chlorite, illite, carbonate mud, and quartz.

Dolostone

Dolostone forms beds ranging from 2 to 75 cm thick and lenses in the form of oblate spheroids with major axes a few meters or less in length and minor axes (perpendicular to bedding) of 75 cm or less. Although not a major portion of any measured section, dolostone beds are laterally much more persistent than graywacke or calcisiltite beds, so they are very useful markers in detailed correlations. Individual dolostone beds have been correlated from sections up to 6 miles apart. Dolostone lenses also tend to occur at the same horizons over comparable distances.

Internal sedimentary structures in the dolostone were not observed in outcrop or in X-radiographs (Hamblin, 1962) of two samples. The boundaries of beds are smooth or, more commonly, gradational into slaty dolomitic argillite. The carbonate grains are uniformly fine (0.002 to 0.02 mm), anhedral, equant, and slightly sutured along contacts, probably as the result of compaction of soft mud. Most of the grains are single crystals; some 2 percent have grown to about 0.06 mm so that they partially enclose several neighboring grains.

One structureless dolostone lens which gradationally replaces part of a massive false-laminated calcisiltite gives some insight into the mass transfer with dolomitization. Relative thicknesses of the lens and the laterally equivalent calcisiltite indicate a volume increase of 115 percent with dolomitization. The dolostone contains 93 percent carbonate (average for bedded dolostones), 3 percent quartz, feldspar, and rock fragments, and 4 percent matrix and carbonaceous material. These figures compare with 25 percent quartz and feldspar and 11 percent matrix and carbonaceous material in the equivalent calcisiltite layer a few inches from the dolostone lens. It appears that the formation of the lens represents a threefold increase in carbonate, including replacement of about 75 percent of the quartz and feldspar by carbonate.

Lenses of dolostone most commonly replace dolomitic argillite; they also occur in the upper, more dolomitic, portions of graywacke beds and as replacements of calcisiltite, particularly in slumped masses. Generally the relations of the lenticular dolostones to adjacent beds indicate that the growth of the lens occurred before appreciable compaction of the replaced argillite and before the underlying few sandstone beds were completely lithified so that they yielded during differential compaction. The surface irregularity produced by the dolostone lens is typically compensated by thinning of the next overlying standstone and does not affect subsequent beds, indicating formation of the dolostone before deposition of the next sandstone bed.

Limestone

Argillaceous limestone beds occur at widely spaced horizons in several members. One massive bed is nearly constant in thickness and appearance for 6 km. The calcite grains are 0.003 to 0.02 mm in diameter, equidimensional, and uniform in size, except for about 3 percent which are recrystallized to 0.06 mm in diameter. The limestones lack sedimentary structures.

Dolomitic Silty Argillite

Distinctive dolomitic argillite beds are found in two members. One member consists almost entirely of a rhythmic repetition of thin calcisiltite beds overlain by dolomitic silty argillite and by normal argillite. No internal or bedding-plane structures were observed in outcrop, in thin section, or in X-radiographs except for planar fabric formed by flattened pieces of carbon and mica seen in thin section. Slaty cleavage is much less apparent in the dolomitic silty argillite than in the adjacent argillite. About 60 percent of the rock consists of illite (identified microscopically). Dolomite occurs as irregular, equant, 0.01-mm grains. Angular quartz grains are 0.01 to 0.06 mm in diameter; elongate or flattened carbonaceous fragments and muscovite flakes are 0.1 mm or less in length. Chlorite, which replaces both matrix and grains, and pyrite are minor constituents.

Volcanic Ash

A few volcanic-ash beds, none thicker than a few centimeters, were found in most units studied in detail. About one-half of the ash is matrix composed of nearly equal amounts of muscovite and montmorillonite and small amounts of pyrite and chlorite. The muscovite flakes, 0.1 to 0.5 mm in maximum diameter, are probably metamorphic. The matrix is compacted or sheathed around coarse silt-sized grains. Broken, rounded, or euhedral crystals of feldspar are the only identifiable grains. Few of the crystals show twinning, but positive signs, optic angles of about 85°, and indices of refraction less than 1.54 suggest compositions of about An_{15}. Many of the grains are only partially devitrified with many small birefringent flecks or vaguely outlined feldspar laths in a murky matrix. Carbonate or sericite partially replace some grains.

Although they are easily obscured, ash beds are useful markers. One bed 1.5 to 2.5 cm thick was traced through five measured sections over a distance of 3 km. A persistent 0.2-cm-thick lamina of carbonaceous montmorillonitic shale occurs 0.5 cm from the top of the bed.

STRATIGRAPHIC UNITS

The Cloridorme Formation is informally subdivided into fourteen members grouped into three sequences called α, β, and γ. Each sequence is confined to one structural block, so the stratigraphic relations between sequences are uncertain. The apparently oldest sequence, consisting of members $\alpha 1$, $\alpha 2$, and $\alpha 3$, is confined to the eastern block and is thrust over the β sequence on the central block. A transverse fault separates the seven members of the β sequence from the four members of the γ sequence on the western block. The offset on the fault and therefore the age relation between the sequences is uncertain. The offset is smaller if the γ sequence is younger, but it is possible that the γ sequence is older and that its four members correspond in part to the α sequence if facies change in the interval between them. The stratigraphic succession and contacts between members are best defined in the β sequence. The α sequence is least known; the contacts are not precisely located, and further subdivision would probably be desirable with more detailed work. The members are exposed along the shore of the St. Lawrence estuary in approximately ascending stratigraphic order from east to west. This is partly because the trend of the gently curving shoreline lies a few degrees northwest of the regional strike. The overturned folds are superposed on a regional dip to the north so that younger rocks are exposed as the shoreline crosses the strike direction westward.

Members are characterized by the proportions of the various lithologic types which they contain and by stratigraphic sequence, as summarized in Table 4. In general, members in which the dominant clastic lithology is graywacke alternate with those in which calcisiltites are dominant. The locations of measured sections are shown in Figure 5 and sections are listed by member in Table 5.

α1 Member

α1 member consists of graywacke, argillite, and type 2 calcisiltite exposed only in a thrust slice at the headland 300 m west of Ruis à Carcy, Jersey Cove (section XXV). The thickness is estimated at 50 m; the mean current direction is 321±23° (Table 5). Because of thrusting, the relations of α1 member to the older Cap-des-Rosiers Formation and other α sequence rocks are unknown.

α2 Member

The type locality of α2 member is Northwest Point, Rivière-au-Renard Ouest (section XXI). The exposed thickness is estimated at 250 m. The member consists of 56 percent argillite, 42 percent calcisiltite, and small amounts of dolostone and calcareous wacke. Although the current directions (294±9°) indicate transport from the southeast, the calcisiltite beds appear to be progressively thicker and to form a larger portion of the member toward the northwest. In section XXI (Figure 5) they average 4.23 cm thick and constitute 32 percent of the rock; in section XIX the figures are 9.85 cm and 40.5 percent; and in section XVIII they are 33 cm and 45 percent. It is unlikely that these short sections were measured at exactly the same stratigraphic interval. The thrust which carries α sequence over the β1 member obscures the base of α2 member. The top is not precisely located either.

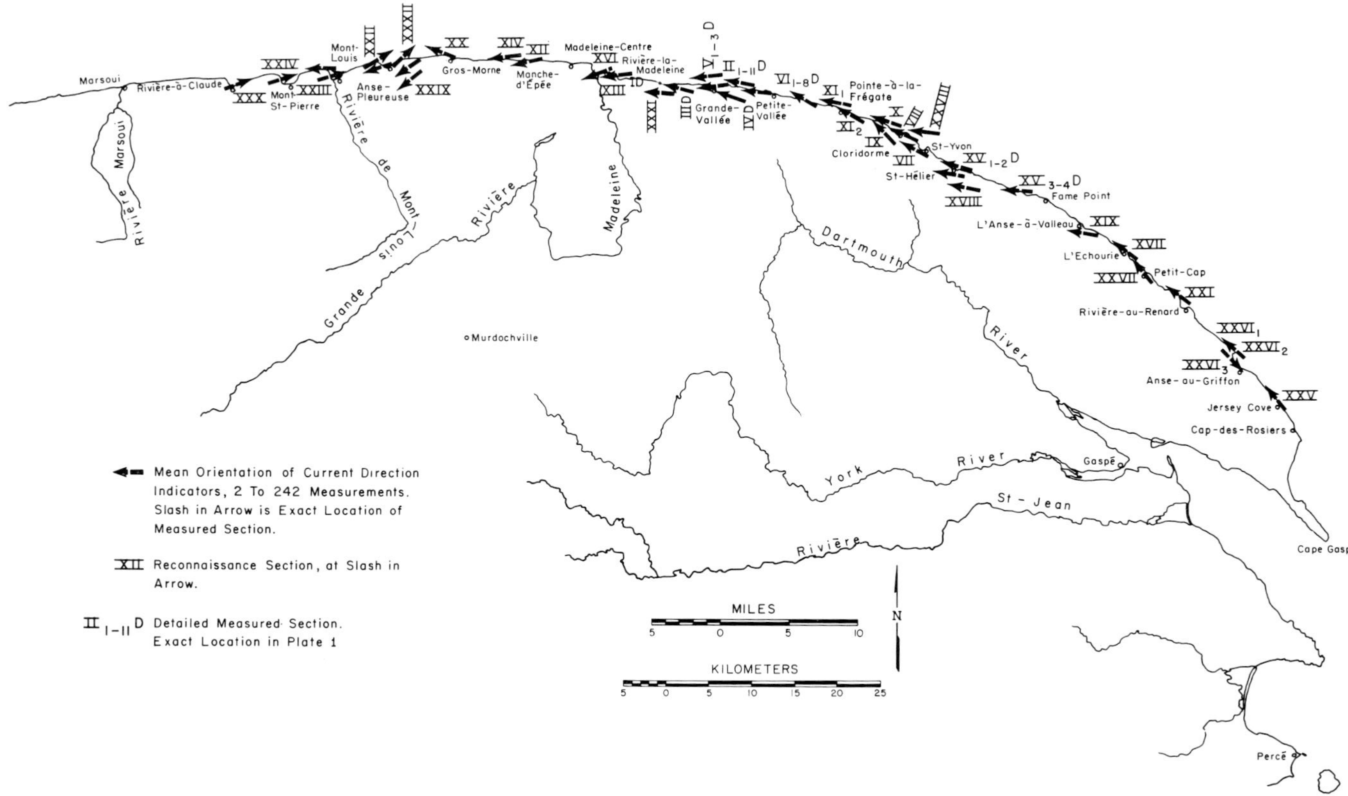

Figure 5. Paleocurrent directions and location of measured sections, Cloridorme Formation (Middle Ordovician), Gaspé Peninsula, Quebec.

α3 Member

Exposures between Cap Serpent at L'Échourie (27 km west of Cap-des-Rosiers, section XVII) and the cove at Petit-Cap (section XXVII) constitute the type of the α3 member. The thickness is estimated at 2000 m, although it may be as much as 3600 m. Argillite, silty dolomitic argillite, type 1 calcisiltite, and calcareous wacke are the principal lithologies, although some massive type 3 graywackes occur in the type section. The rocks at Anse-au-Griffon and west are assigned to this member, although they contain 70 percent calcareous argillite and 5 percent silty limestone. They are probably higher in the section than the type locality. Current directions are 313±8°, except for a portion of the section at Anse-au-Griffon with directions of 138±22°. The α3 member probably overlies the α2 member east of Rivière-au-Renard in normal sequence.

β1 Member

The cove at St.-Hélier (8 km east of Cloridorme) is the type locality of β1 member, which is about 1175 m thick and is characterized by type 3 graywackes that are found elsewhere in abundance only in γ1 member. Calcareous wacke, type 1 calcisiltites, dolostone, and argillite complete a distinctive assemblage of lithologies. The mean direction of depositing currents was 282±15°. The base of β1 member is everywhere covered by the major thrust which bounds the central and eastern blocks, except perhaps at Pointe-Jaune where calcisiltites which may belong to α3 member are exposed. The top of the member, placed at the highest type 3 graywacke bed, is exposed 50 m north of the pier at Pointe de Cloridorme, Cloridorme-Ouest; along Route 6 on the hill at the eastern edge of Pointe-à-la-Frégate; and in the sea cliffs 1.9 km west of St.-Hélier, the type locality.

β2 Member

β2 member is exposed between Cloridorme-Ouest, the type locality, and Cap Barré, 1.8 km east of Cloridorme. The member is 490 m thick and consists of argillite, thin-bedded calcisiltite, and calcareous wacke which becomes less abundant near the base. Current directions average 299±22°. The base is exposed at the type locality and other localities mentioned above. The upper contact with β3 graywackes is exposed 500 m south of Cap Barré and at Anse aux Canons, Pointe-à-la-Frégate.

β3 Member

The entire thickness of β3 member is fairly well exposed south of Cap Barré near St.-Yvon, the type locality, and at the eastern edge of Pointe-à-la-

Frégate. It is 43 m thick at Cap Barrè, although it apparently thins westward. Argillite, calcisiltite, and graywacke are the principal rocks. Both contacts are well defined by the first occurrence of graywackes, which form groups of fairly thick beds. No current directions were measured.

β4 Member

Section VII at St.-Yvon includes the entire thickness of β4 member, 78.5 m, and is the type locality. Type 1 calcisiltite and argillite are the dominant rocks. Calcareous wackes are nearly absent in section VII, but they form 13.5 percent of the top 10 m of the member in Section XI at Pointe-à-la-Frégate, although it is downcurrent from section VII. Mean current direction is 304±25°. Contacts with the graywackes which overlie and underlie β4 member are exposed at St.-Yvon and Pointe-à-la-Frégate.

β5 Member

The wave-cut benches at Point-à-la-Frégate (section XI) are the type locality of β5 member, consisting of about 80 m of graywacke, calcisiltite, and argillite. The orientation of current-direction indicators is variable; the mean is 287±32°. The base of the member is marked by an abrupt change from calcisiltite to dominantly graywacke as exposed at Pointe-à-la-Frégate, in the syncline at St.-Yvon (section VII), and east of Pointe Sèche, St.-Yvon. The top, which is drawn at the highest 10 graywackes within a 10-m interval, is exposed in the wave-cut benches west of Pointe-à-la-Frégate, but strike-slip faults and covered intervals separate these exposures from section XI and from section VI_8 of β6 member.

β6 Member

The narrow wave-cut bench 4.5 km west of Pointe-à-la-Frégate (section VI_8) exposes the entire thickness of β6 member except for two small covered intervals. It is about 255 m thick at this, the type locality, and is characterized by type 1 calcisiltite and dolostone. Argillite comprises 78.5 percent of the rock, the highest pure argillite content of any member of the Cloridorme Formation. The mean current direction is 295 ±15°.

The base of the member, which is exposed in fault slices at the base of section VI_8 west of Pointe-à-la-Frégate, is drawn at the top of the highest 10 graywackes within a 10-m interval. Some interlayering of the graywacke and calcisiltite units exists at the contact. The top of the member, exposed in sections VI_8, VI_{3-4}, and II_3, is drawn at the base of the lowest 10 graywackes within a 10-m interval, which happens to include the lowest graywacke. A transition zone 130 m thick is thus placed entirely in β7 member. This transition

TABLE 5. SUMMARY OF PALEOCURRENT MEASUREMENTS

Member	Section	Location	Orientation	Standard Deviation±	Number Readings
γ4	XXX	E. edge Rivière-à-Claude	70.4°	13.65°	17
γ3	XXXI	Route 6, sharp curve E. Ruis la Rouge	(276.4°)	(7.96°)	(7)
	XXXII	Wave-cut bench, 150 m E. Anse-Pleureuse	45.2°	12.89°	5
	(XXXII)	W.-directed currents only	(230.5°)	(15.49°)	(11)
	XXII†	W. side, Anse Pleureuse	58.3°	14.69°	4
	(XXII)	W.-directed currents only	(260.7°)	(31.90°)	(3)
	XXIII	Pier, Anse de Mont-Louis	73.2°	7.01°	45
	(XXIII)	W.-directed currents only	(271.5°)	(12.02°)	(2)
	XXIV	W. side, Baie de la Rivière-à-Pierre, Mont-St.-Pierre**	72.5°	13.79°	29
	Total	(W.-directed currents only)	(252.0°)	(16.16°)	(23)
		E.-directed currents	70.5°	10.55°	83
γ2	XIII	Black Point, Rivière-la-Madeleine**	263.7°	20.01°	132
	XVI	Cap à l'Ours, Rivière-la-Madeleine	242.2°	28.04°	23
	XX	Cap du Gros-Mâle, Gros-Morne	291.5°	40.03°	51
	XXII†	W. side, Anse Pleureuse	250.4°	23.99°	42
	XXIX	Cliff, SE. end Lac de l'Anse-Pleueuse	231.6°	15.08°	18
	Total		262.0°	26.22°	266
γ1	XII	E. side Manche-d'Épée**	258.7°	13.31°	76
	XIV	Coast W. Manche-d'Épée**	263.0°	13.17°	38
	Total		260.1°	13.26°	114
β7	II_9	Municipal dump, 2 mi. E. Grande-Vallée	276.4°	31.17°	112
	II_8	600 m. E. Anse à Mercier, Grande-Vallée	282.4°	32.81°	90
	II_4	E. Anse à Colin, Grande-Vallée	265.9°	34.62°	248
	V_{1-3}	Cap Barré, Grande-Vallée	264.3°	23.15°	44
	III_{4-5}†	Route 6, E. Le Grande-Ruisseau	287.5°	16.84°	14
	VI_8†	1000 m E. Pointe Ronde**	293.9°	27.26°	98
	Total		275.2°	31.62°	606

$\beta6$	VI_8†	Headland 1000 m. E. Pointe Ronde**	302.5°	13.48°	31
	$III_{1\text{-}3}$†	Route 6, E. Le Grand-Ruisseau	298.5°	12.11°	6
	IV	Grande-Vallée Harbour, S. Route 6	290.0°	16.41°	50
	Total		295.3°	15.19°	87
$\beta5$	VII†	W. side, Baie St.-Yvon	291.8°	27.75°	35
	XI†	W. headland, Pointe-à-la-Frégate**	283.9°	33.21°	82
	Total		286.7°	31.68°	117
$\beta4$	VII†	W. side, Baie St.-Yvon**	305.7°	25.08°	57
	X†	E. headland at Pointe-à-la-Frégate	288.9°	10.61°	4
	Total		304.4°	24.55°	61
$\beta2$	VIII	W. harbour, Cloridorme**	295.6°	24.21°	122
	IX	W. Baie de Cloridorme, Cloridorme-Ouest**	317.4°	26.55°	60
	X†	Pointe de Cloridorme**	302.6°	24.59°	4
	XI_2	Anse aux Canons, Pointe-à-la-Frégate	300.3°	13.01°	40
	XXVIII	Cap Barré, W. St.-Yvon	277.8°	6.92°	33
	Total		298.9°	21.96°	259
$\beta1$	XV_3	Ruisseau à l'Échalote, W. of Fame Point	274.2°	9.23°	114
	XV_2	E. side Anse de l'Étang, St.-Hélier**	286.9°	12.59°	182
	XV_1	E. side Anse de l'Étang, St.-Hélier**	281.4°	17.17°	242
	X†	Pointe de Cloridorme	285.2°	18.92°	15
	Total		281.77°	14.51°	553
$\alpha3$	XVII	N. of Church St.-Maurice L'Échourie**	307.8°	6.29°	52
	$XXVI_{1\text{-}3}$	English Point, Anse au Griffon	314.5°	13.54°	19
	$(XXVI_{1\text{-}3})$	E.-directed currents only	(138.3°)	(21.81°)	(14)
	XXVII	200 m. S. Cap Barré, Petit-Cap**	321.2°	4.61°	30
	Total		313.1°	7.78°	101
$\alpha2$	XVIII	S. end Grand Étang	292.4°	5.98°	7
	XIX	Route 6, L'Anse-à-Valleau	280.4°	6.97°	27
	XXI	Northwest Point, Rivière-au-Renard-Ouest**	307.6°	11.09°	29
	Total		294.2°	9.06°	63
$\alpha1$	XXV	W. Ruis à Carcy, Jersey Cove**	320.9°	23.29°	28

**Denotes section of type locality.
†Section includes part of adjacent member also.

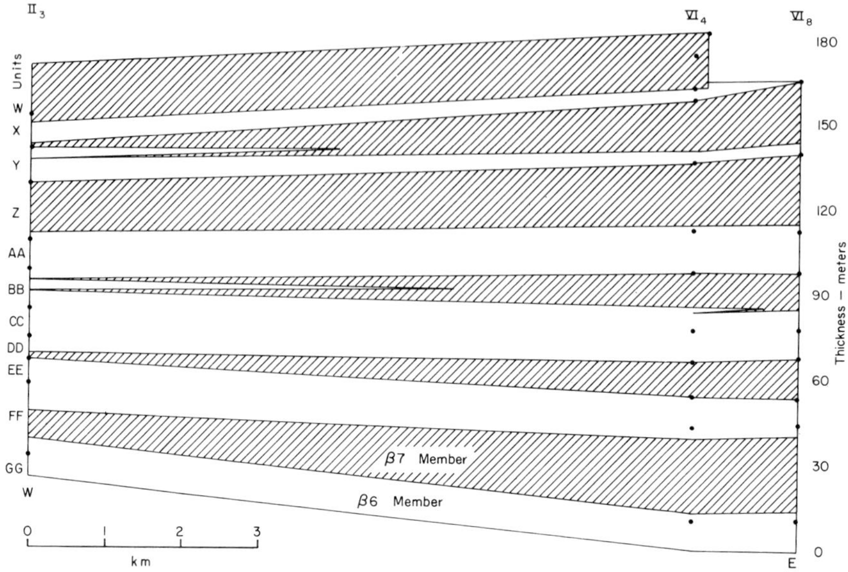

Figure 6. Intertonguing of graywacke-argillite intervals (pattern) and calcisiltite-argillite intervals (blank) at the base of $\beta 7$ member, sections II_3, VI_4, and VI_8. Mean current directions are from the east (right). Dots show positions of selected key beds.

zone is formed by intertonguing of predominantly graywacke units and calcisiltite units (Fig. 6) as graywacke gains ascendency to the east, the upcurrent direction. Fairly persistent dolostone beds provide stratigraphic control, indicating that the top of $\beta 6$ member rises toward the west, although the boundary is defined so that this effect is only a few meters in the 10 km between measured sections VI_8 and II_3. Section III_3, which is 80 m thick and contains only 3 per cent graywacke in road cuts 4.4 km west of Grande-Valée (section II_3) may represent the westward continuation of this intertonguing with further disappearance of graywackes, but small thrust faults which terminate section III_3 make its exact stratigraphic position uncertain.

$\beta 7$ Member

The type locality of $\beta 7$ member is the coastal exposures between Petite-Vallée and Pointe Ronde, 2.2 km east. Graywacke is the dominant clastic rock in the 835 m exposed. The mean direction of depositing currents was 275±32°. The base of $\beta 7$ member is exposed at Grande-Vallée, at Pointe Ronde, and in section VI_8 east of Pointe Ronde. The top is not exposed because of the shoreline and a syncline to the north, the transverse fault which limits the exposure to the west, and the lack of satisfactory inland exposures.

γ1 Member

Manche-d'Épée (21.4 km east of Mont-Louis) is the type locality of γ1 member; at least 550 m are exposed in the narrow wave-cut benches and spectacular seacliffs between Madeleine-Centre and Gros-Morne. Dolomitic silty argillite comprises 57 percent of this member, the only one in which it is common. Type 3 graywacke and dolomitic argillite (14 percent of the member) are other principal rocks. Current directions, derived mainly from the well-exposed bottom part of the member, average 260±13°. The contact of γ1 member with graywackes of γ2 member is exposed in the overturned sequence in the sea cliff capped by the church of Gros-Morne. The base of γ1 member is not exposed.

γ2 Member

The type locality of γ2 member is the headlands enclosing La Grande Anse, 11 km west of Grande-Vallée (section XIII). Graywacke and massive type 2 calcisiltite characterize the 800-m-thick member. The mean orientation of directional structures is 262±26°. The base of γ2 member is exposed beneath the church at Gros-Morne in sharp contact with dolomitic silty argillite of γ1 member and near the southern end of Lac de l'Anse-Pleureuse in section XXIX. The top of the member is a rapid transition from graywacke to thin calcisiltite of γ3 member, exposed, despite some complications by faulting, in the wave-cut platform on the western side of Anse-Pleureuse (section XXII).

γ3 Member

The cliffs west of the bay at Mont-St.-Pierre (section XXIV) are the type locality of γ3 member, estimated to be 475 m thick. The principal lithologies are argillite, dolomitic argillite (18 percent of member), calcisiltite, calcareous wacke, and dolostone; relative abundances vary noticeably within the member. Abundant dolomitic argillite and at least some silty dolomitic argillite occur at the basal contact. Section XXIII in the lower half of the member at Mont-Louis contains 26 percent dolomitic argillite, 5.5 percent dolostone, and 12 percent calcareous wacke. Calcisiltite is the only important nonpelagic lithology in the upper part of the member; it comprises 21 percent of the Mont-St.-Pierre section. Most directional structures are oriented eastward; these average 70±11°, or longitudinal to the trough as in the lower members but originating from the opposite direction. About one-fifth of the currents still flowed in the opposite direction, 252±16°. Beds with opposing paleocurrent directions are interlayered at the base of the member. The sections are arranged in order from east to west in Table 5 to demonstrate that beds deposited by west-directed currents disappear toward the west and *vice-versa.*

The section at Mont-Louis and those near Anse-Pleureuse are of roughly equivalent rocks near the base of the member.

The base of the member is seen at Anse-Pleureuse and the top was approximately located at Petit Cap, 10 km west of Mont-Louis, where massive graywacke of $\gamma 4$ member appears abruptly in the section.

$\gamma 4$ Member

Graywacke of $\gamma 4$ member crops out from Petit Cap, near Mont-St.-Pierre, westward to Marsoui. The thickness exposed is estimated at 510 m. About 60 percent of the member is massive beds of graywacke; the remainder is calcisiltite and argillite. The graywacke and calcisiltite percentage combined is about 80 at the lower contact, 40 slightly higher in the member (section XXX), 65 at the middle, and 80 in the upper half. This abundance of graywacke may have led to confusion of these rocks with the massive "Pillar sandstone" of the Cap-des-Rosiers Formation in the Tourelle area east of Ste.-Anne-des-Monts (McGerrigle, 1954b), but the graptolites from the two units indicate that their times of deposition were separated by several stages (p. 40; McGerrigle, 1954b, p. 29; Berry, 1962b, p. 710). The current-direction indicators average $70 \pm 14°$, showing transport from the west as in $\gamma 3$ member. These graywackes, alone among the Cloridorme Formation rocks examined, contain no serpentine or chromite, and quartz is more abundant (64 percent); an important difference in source areas as well as transport direction is indicated.

An abrupt change from calcisiltite deposition to graywacke deposition evident at Petit Cap marks the base of $\gamma 4$ member. An upper contact is not exposed; outcrops of Cap-des-Rosiers rocks west of Marsoui (p. 15) indicate that a structural dislocation of great magnitude truncates the member to the west.

AGE AND CORRELATION

The age of the Cloridorme Formation has long been recognized as Middle Ordovician and closely related to the Normanskill Formation of New York (Jones, 1934a; McGerrigle, 1954a, 1959), which is assigned by Berry (1962b) to the Wilderness and early Trenton stages of Cooper's (1956) classification. Collections of graptolites, mainly from $\beta 7$, $\beta 5$, and $\beta 1$ members, contain many individuals but surprisingly few species, considerably fewer than in McGerrigle's collections. *Orthograptus truncatus* varieties (all identifications and age information are courtesy of Dr. W. B. N. Berry) are by far the most abundant forms in the α and β sequences, followed by

Climacograptus spiniferus. Climacograptus spiniferus and *Orthograptus truncatus* var. *pertenuis* are restricted to rocks of Trenton age (in the sense of Cooper, 1956) in North America and *O. truncatus* var. *intermedius* is most common in rocks of Trenton age. All of the Cloridorme Formation exposed on the northern Gaspé coast apparently belongs to the *Orthograptus truncatus intermedius* zone (Berry, 1960) of late Wilderness and Trenton age. Only the uppermost Normanskill in New York (upper part of Austin Glen Member; Berry, 1962b) is of this age, so the Cloridorme Formation may correlate with the youngest Normanskill or the overlying Canajoharie Formation (Berry, 1964, written commun.).

The graptolite fauna by members follows:

$\alpha 2$ member, section XIX, L'Anse-à-Valleau

- *Climacograptus* n. sp.
- *Diplograptus* sp.
- *Orthograptus truncatus* cf. var. *pertenuis* (Ruedemann)
- *Orthograptus* sp. (of the *O. truncatus* group)

$\beta 1$ member, section XV_2, St.-Hélier

- *Dicranograptus* sp.
- *Orthograptus* sp. (of the *O. truncatus* group)

$\beta 2$ member, Route 6 road cut at St.-Yvon

- *Climacograptus* sp.
- *Diplograptus* sp.

$\beta 5$ member, section XI, Point-à-la-Frégate

- *Climacograptus* sp.
- *Orthograptus* sp. (of the *O. truncatus* group)

$\beta 6$ member, section II_1, Grande-Vallée

- *Lasiograptus* sp.

$\beta 7$ member, largest collection from section II_3 Grande-Vallée. Also from section II_4, Grande-Vallée, and section III, Route 6 road cut eastern side of Le Grand-Ruisseau, 4.5 km west of Grande-Vallée

- *Climacograptus brevis* Elles and Wood?
- *Climacograptus spiniferus* Ruedemann
- *Climacograptus* sp.
- *Diplograptus* sp.
- *Orthograptus truncatus* cf. var. *pertenuis* (Ruedemann)
- *Orthograptus truncatus* cf. var. *intermedius* Elles and Wood
- *Orthograptus* cf. *O. truncatus* (Lapworth)
- *Orthograptus* sp. (of the *O. truncatus* type)

$\gamma 2$ member, section XX, cape 1 km west of Gros-Morne, and section I, cliff east of La Grande Anse, 11 km west of Grande-Vallée

Climacograptus scharenbergi Lapworth?
Dicranograptus sp.
Diplograptus sp.
Orthograptus sp. (of the *O. truncatus* group)
Orthograptus truncatus cf. var. *intermedius* Elles and Wood

$\gamma 4$ member, Route 6 road cut, 2.7 km east of Marsoui

Climacograptus spiniferus Ruedemann
Diplograptus sp.

Nearly all of the graptolites were collected from the graywackes where they are preserved as a carbonaceous residue that has not been flattened as in the shales but which preserves little detail of the organisms. Finds of particular interest taxonomically are four synrhabdosomes (colonial groups) of orthograptids ($\beta 7$ member, section VI_3) and some unusually large dicranograptids with stipes up to 22 cm long in good three-dimensional preservation ($\beta 1$ member, section XV_2).

The reworked brachiopods and other fossils are too fragmented to be of use in age determination. Nautiloid cephalopods from the argillite belong to the orders Endoceratida and Michelinoceratida, perhaps of Denmark or Cobourg age (Trenton; R. H. Flower, 1964, written commun.).

Flysch of the age of the Cloridorme Formation occupies a narrow trough in the valley-and-ridge folded belt for much of the length of the Appalachians (Fig. 2). On Newfoundland the younger rocks of the Humber Arm Series (Schuchert and Dunbar, 1934) are of the same age as the Cloridorme Formation. Middle Ordovician flysch is known from a few localities along the St. Lawrence: L'Islet, Charny, and Quebec City (Claude Hubert, 1966, personal commun.; Osborne, 1956). The Quebec City Formation and older rocks of the Quebec Group extend south through the St. Lawrence lowland and interfinger with the miogeosynclinal Champlain Valley sequence at the international boundary (Cady, 1960), but the flysch belt continues through the eastern townships of Quebec in the form of the Middle Ordovician Beauceville Formation (Cooke, 1950). The Beauceville or Magog rocks are equivalent to metamorphosed rocks of the Cram Hill Formation east of the Green Mountain axis in Vermont (Currier and Jahns, 1941; Berry, 1962a). These rocks are lost in higher grade metamorphic terranes to the south and are eroded from the Green Mountain axis, but the thick flysch of the Taconic range to the west includes the Pawlet Formation, an approximate equivalent of the Normanskill (Zen, 1961). Upper Middle Ordovician and older flysch deposits are extensively exposed in the Hudson Valley of New York, the type locality of the Normanskill Formation (Ruedemann, 1901; Berry, 1962b). From southern New York, through the great valley of Pennsylvania and south as far as eastern Tennessee, the Martinsburg Formation is at least partially age equivalent and facies equivalent to the Normanskill Formation (McBride, 1962; Rodgers, 1957).

Black shale and sandstone of the Canajoharie and Utica Formations, which overlie the Normanskill in the Hudson Valley, extend along the Mohawk Valley into western New York, far beyond the former narrow confines of the flysch trough. Similar relations are found in the Martinsburg Formation and the partially equivalent Reedsville Formation to the west from Pennsylvania to Tennessee. This relation may also exist in Gaspé Peninsula between the Cloridorme Formation and the Makasti Shale (Twenhofel, 1928), an upper Middle Ordovician black shale named from blocks washed onto the north shore of Anticosti Island apparently from Jacques Cartier Passage. This black shale unit extends to the edge of the Canadian Shield between two limestone sequences of the continental platform.

Across strike to the southeast, the Mictaw Formation of southeastern Gaspé (Ayrton, 1961; Alcock, 1935) is somewhat similar in lithology to the Cloridorme Formation and is probably late Middle Ordovician in age, but it contains no graptolite species in common with the Cloridorme Formation, or the type Normanskill Formation. It may be somewhat younger (Ruedemann, *in* Alcock, 1935, p. 14). The Tetagouche Series of northern New Brunswick (Alcock, 1935, 1941) includes some flysch containing many graptolite species in common with that part of the type Normanskill older than the Cloridorme Formation.

ORIGIN

The origin of the Cloridorme Formation may be considered in terms of transporting mechanisms, environment of deposition, and source area.

Transporting Mechanisms

The origin of sequences of fine-grained pelagic mud with interlayered coarse-grained terrigenous material, such as the Cloridorme Formation, was a problem which defied adequate explanation until the concept of turbidity currents was applied as a transporting mechanism (Kuenen and Migliorini, 1950; Shepard, 1951; Bell, 1942). A characteristic suite of sedimentary structures was recognized from the study of exposed sandstones whose deposition is best explained by turbidity currents (Kuenen and Migliorini, 1950; Rich, 1950; Crowell, 1955; Kuenen, 1957). These include most of the structures found in the calcisiltite, calcareous wacke, and graywacke of the

Cloridorme Formation: flutes, grooves, prod marks, bounce marks, skip marks, brush marks, chevron marks, oriented clasts and fossils, convoluted lamination, current ripple lamination, load flowage structures, shale fragments, and graded beds.

None of these structures is a foolproof criterion of turbidity-current action, or at least of turbidity currents in deep submarine basins, because most of them have been reported from vastly different environments. For example, flutes have been reported from fresh-water deposits (Cummins, 1958; Belt, 1963) and shallow-water Paleozoic sandstones of the midcontinent (Potter and Glass, 1958, Pl. 4B; Potter and Pettijohn, 1963, Pl. 13B). Load flowage structures occur in many environments, including lagoons (Emery, 1950) and stream deposits (Kaye and Power, 1954). Graded beds are common in volcanic-ash deposits and are known from stream deposits and from fresh-water lakes (Sanders, 1965). Potter and Glass (1958, p. 18-23) have described flutes, load flowage structures ("load casts"), prod marks (Pl. 3B), convoluted lamination, oversteepened cross-bedding, and slump structures from shallow marine and continental clastic rocks of the Pennsylvanian in Illinois, although none are abundant and some are of dimensions atypical of flysch. Examples of characteristic turbidite sedimentary structures from nonflysch, and in some cases nonturbidite, sequences could be multiplied. The great abundance and variety of characteristic structures found in the Cloridorme Formation, however, constitute numerous lines of evidence converging to make a much stronger, though not absolutely conclusive, case.

Strong evidence for resedimentation by turbidity currents is the presence of reworked shallow-water fossils in a deep-water environment (Natland and Kuenen, 1951; Ericson and others, 1952; Ksiazkiewicz, 1952) The only fossils found in the pelagic argillite of the Cloridorme Formation are planktonic graptolites and nektonic nautiloids. No evidence of a bottom fauna is found except some questionable burrows and extremely rare feeding marks. Fragmented and size-graded fossils of benthonic organisms such as brachiopods, bryozoans, gastropods, corals, and echinoderms are abundant in many sandstone beds. Thus in the Cloridorme Formation transported benthonic forms, some of which were probably shallow-water dwellers, are mixed with pelagic forms which were independent of depth and bottom conditions.

The evidence for transportation and deposition by turbidity currents may be summarized for each rock type. Negative evidence applicable to each lithology is the apparent lack of wave action or subaerial exposure, indicating that deposition of the coarse clastic material occurred below wave base.

Calcisiltites of type 1 contain flutes, prod marks, bounce marks, grooves, and a few benthonic fossils. Ripple-drift cross-lamination indicates vigorous current action with rapid fallout of detritus. Convolute lamination is most common in turbidites (ten Haaf, 1956; Sanders, 1960). The thin layers of calcisiltite abundant in $\beta 7$ graywacke member and elsewhere may result from normal bottom currents carrying some silt, perhaps the "steady turbidity currents" of Dunbar and Rodgers (1957, p. 14) or other density currents, as many of them lack sedimentary structures or are cross-laminated. However, a few contain convoluted laminations or flutes.

Discontinuous lenses of siltstone, some of which appear to be isolated ripples, are found locally. These resemble Hubert's (1964) description of silty layers from the North Atlantic abyssal plains which he concluded were deposited by traction currents other than turbidity currents. As far as can be ascertained, however, a continuum exists between these lenses and type 1 calcisiltites attributed to turbidity-current deposition. At present, definitive criteria are lacking to distinguish between these two depositional mechanisms in most individual beds, ancient or Recent.

Despite petrologic similarities between type 1 and type 2 calcisiltites, differences in texture and sedimentary structures indicate differences in mode of transport and deposition. Possible transport mechanisms appear to be "normal" submarine currents and turbidity currents, spasmodic or steady (Dunbar and Rodgers, 1957, p. 14). The greater thickness of type 2 calcisiltites suggests currents of greater capacity, but the lateral persistence of type 2 beds is so low that the total volume of sediment in them may be no more than in type 1 calcisiltites if these are much more laterally persistent as suggested by correlations. The paucity of flutes, tool marks, and convoluted lamination may reflect relatively slow currents. The greater content of fine-grained matrix in type 2 calcisiltites and their lateral gradations into shale need not involve slower currents to account for the fine material; on the contrary, more rapid currents involving erosion of the bottom might be assumed except for the absence of flutes (scour marks). Lack of flutes could reflect the presence of a traction carpet protecting the bottom rather than a slow current (Dzulynski and Sanders, 1962, p. 86 *ff.*), but the grain size in type 2 calcisiltites is equal to or less than in type 1 calcisiltites; thus the presence of a traction carpet only in type 2 would in itself imply slower currents. Climbing ripple-drift cross-laminae require the constant addition of material by fallout from suspension during deposition; this amount of material being carried in suspension means that regardless of the origin of the depositing current, it was certainly turbid at the point of deposition. The transporting agent most consistent with these interpretations and the short lateral extent of the beds seems to be spasmodic turbidity currents of lower velocity and less turbulence than those which deposited type 1 calcisiltites. The slower velocity would lead to more rapid deposition of thicker beds with small lateral persistence without necessitating currents of great volume. The problem then is maintaining these slower currents into the area where they deposited rapidly; proximity to the submarine slope down which the currents moved may be necessary.

Flutes, a profusion of small tool marks, graded beds convoluted lamination, and abraded benthonic fossils are found in calcareous wackes. Several seemingly anomalous features are associated in these rocks. The presence of flutes on more than half of the beds suggests that more erosion of the bottom mud was involved in the transportation of calcareous wackes than of coarse graywackes, but argillite and calcisiltite fragments are rare and the matrix content is significantly less. The argillaceous upper portions of some beds may be the result of bottom erosion, but, if so, the degree of separation of silt and clay fractions is surprising. The absence of large fragments of argillite and calcisiltite, except in the upper parts of some beds, explains the absence of grooves by the lack of tools. The presence of flutes is thought by Dzulynski and Sanders (1962, p. 86 *ff.*) to depend upon the absence of a traction carpet to protect the bottom mud from eddies, but flutes here underlie the thin-to-lenticular ripple cross-laminated layers interpreted as a traction carpet (Fig. 4).

Graded beds of coarse sand interbedded with pelagic shale suggest turbidity-current deposition of the graywackes. All three types contain abraded fragments of benthonic brachiopods, bryozoans, corals, and pelecypods and structures such as flutes, grooves, argillite and calcisiltite fragments, and slump structures.

The remaining fine-grained rocks—dolostone, limestone, dolomitic silty argillite, and part of the argillite—are ascribed to pelagic accumulation. Argillite in the Cloridorme Formation is considered the "background" sedimentation, the material which rained down at a relatively constant rate from suspension in the overlying water. Most beds other than argillite represent short-lived events superimposed on the background, events such as the influx of a turbidity current. Much of the argillite is deposited by turbidity currents, however, which considerably muddies this picture of delightful order and simplicity. Graywacke beds can be traced laterally into argillite beds of comparable thickness. The amount of argillite in most units decreases significantly downcurrent; the rate of thinning relative to that of other lithologies indicates that about half the argillite is of turbidity-current origin (Enos, 1969). Thus conclusions about rates of pelagic deposition, correlations based on thickness, and interpretations of sandstone/shale ratios are made more complicated. Further complications arise from the erosion of the pelagic mud by turbidity currents.

No internal structures and sole marks are found to indicate current action in the origin of dolostone and limestone beds. These beds are much more persistent laterally than are graywackes or calcisiltites and many of them grade at top and base through dolomitic or calcareous argillite into pelagic argillite. It is inferred that they originated as pelagic oozes or precipitates. What conditions led to deposition of limestone rather than argillite and of dolostone rather than limestone is an unsolved mystery. The rate of deposition of dolostone appears to have been about 14 times that of the argillite and the limestone about 6 times that of the argillite, based on the percentage of pelagic noncarbonate mud in each lithology and the assumption that its deposition rate was constant but its percentage diluted by carbonate deposition.

The origin of the strikingly massive, homogeneous, and resistant beds of dolomitic silty argillite is enigmatic. The only suggestion that they are other than a pelagic deposit is that more than 90 percent of the beds overlie thin calcisiltite beds, presumed to be turbidity-current deposits, and that the thickness of argillite beds is positively correlated with the thickness of the turbidite beds. A genetic relation is strongly suggested, but, in the absence of any structures which indicate current action in deposition of the argillite, it is possible that the relation is casual or perhaps causal. For example, a storm or other event which triggered the turbidity current may have mixed carbonate and terrigenous mud into the surface water layers whence it settled out as a pelagic deposit.

Depositional Environment

The environment of deposition of the Cloridorme Formation may be inferred from the depositional or early diagenetic mineral assemblages and from fossilized autochthonous fauna or the lack thereof. The role of the dolostone, limestone, and part of the argillite as slow, pelagic accumulations makes them the most valuable indicators of depositional environment, for they are the sediments most likely to have been at equilibrium with local bottom conditions.

The mineral assemblage includes pyrite and calcite. Pyrite and organic matter were present as finely divided constituents of the bottom mud and are found in all the rock types. Calcite is not normally present in the argillite, but there are numerous gradations of the argillite into calcareous and dolomitic argillite. Thus it appears that the *p*H must have lain near stability for limestone deposition, about 7.8 (Krumbein and Garrels, 1952) during the deposition of these rocks. Calcite or dolomite is present in all other lithologic types, but most or all of this carbonate was introduced as lime mud and fossils from outside the immediate environment.

The association of calcite with pyrite and organic matter as stable phases suggests that the Eh (redox potential) was less than −2.5. That this Eh obtained at some time apparently early in the history of these rocks does not answer the question of whether this represents the condition of the bottom waters during deposition or a condition found only in the mud at or below the depositional interface. The best evidence for determining this is faunal. The rare fossils found in the argillite are all pelagic forms, cephalopods and graptolites. The only indication that enough oxygen to support life was available in the bottom water is from trace fossils. Questionable feeding marks, resembling *Palaeodictyon* (Menegh.), figured by Seilacher (1959, p. 1068, Pl. 4B), are found on the base of a few graywacke beds in $\alpha 1$ member. Numerous small knobs on the base of many calcisiltite beds may be fillings of the tops of burrows in the underlying mud. If the organisms which made these traces were aerobic, the bottom water must have been somewhat aerated and negative redox potentials must have been confined to the bottom mud. The *p*H values of greater than 7.8, on the other hand, reflect the periodic condition of the water from which the limestone or dolostone was deposited. Carbonate remained a stable phase relative to quartz and feldspar during diagenesis of the nonpelagic rocks as the abundance of embayed or replaced terrigenous grains demonstrates.

The framboidal texture of most of the pyrite suggests that it is authentic (Love, 1962a, 1962b). The dimpled pyrite spheres, 6 to 12 microns in diameter, are similar to those found in Recent marine and lake muds and in fine-grained sedimentary rocks of all ages (Love, 1962b). They appear to be products of very early organic diagenesis in anaerobic environments (Vallentyne, 1962; Love 1962b). Larger spherical or irregular aggregates (average 0.3 mm) of the framboidal spheres are present in the graywackes, a lithologic association not previously reported, in the other nonpelagic rocks, and in the dolostones. It is likely that they are the form of much of the disseminated pyrite in the argillite.

Source Area

The mineralogy of the resedimented lithologies—graywacke, calcisiltite, and calcareous wacke—was studied for clues about the source area. The pelagic deposits were transported as very fine suspended particles or in solution and so yield little information about their source.

The heavy-mineral suite of the resedimented rocks is rather undiagnostic with one notable exception—chromite. Octahedral or broken grains, 0.005

to 0.1 mm in diameter, of a slightly transparent, reddish-brown isotropic mineral with a refractive index between 2.026 and 2.029 and an isometric unit cell of 8.24 Å were present in nearly all separations and were the dominant nonopaque mineral in some. This mineral is chromite with a composition of $(Fe_{0.6}\ Mg_{0.4})\ (Cr_{0.64},\ Fe_{0.06},\ Al_{0.3})_2 0_4$ in mole proportions calculated from the regression formulas of MacGreggor and Smith (1963). Some range in composition is indicated by variation in color and refractive index; light-brown, transparent grains have refractive indices between 1.926 and 1.937. Chromite is almost exclusively associated with ultramafic intrusives such as the Mt. Albert complex of the Shickshock Mountains (Fig. 1) 35 km south-southwest of Marsoui (MacGreggor and Smith, 1963), the Mt. Serpentine intrusive 27 km south of Cloridorme (Jones, 1934b), and the serpentine belt of western Newfoundland (Smith, 1958). The mafic intrusives, largely anorthosites, of the Canadian Shield north of the St. Lawrence estuary do not contain appreciable chromite (I.D. MacGreggor, 1963, written commun.).

Other heavy minerals include abundant zircon, some tourmaline, and traces of apatite, garnet, hornblende, and magnetite in graywackes and zircon, tourmaline, and a trace of rutile in calcisiltites. Pyrite is abundant, but most of it appears to be authigenic. This assemblage is distinctive in its maturity—only very stable and resistant grains are found. This could reflect extremely deep weathering in the source area or considerable recycling, or both; recycling is favored because of the presence of relatively unstable rock fragments which could not survive prolonged weathering and of some well-rounded and metamictized zircon.

The quartz grains are generally subangular with an estimated rounding index of 0.35 (Krumbein, 1941), although irregular boundaries resulting from diagenetic solution and replacement at the borders by calcite is extensive and may have influenced the estimate. Nevertheless, many grains of medium sand size are well rounded, which requires numerous sedimentary cycles according to the experimental data of Kuenen (1959) on fluviatile abrasion. Much of the feldspar is altered and corroded to the point of uselessness in determining composition or even roundness. Although albite twinning is rare, cobaltinitrite staining (Bailey and Irwin, 1959) shows that virtually all the feldspar is plagioclase except for rare grains of microcline. Plagioclase grains in which the composition could be determined were oligoclase.

The most abundant rock fragments (Table 1) are of volcanic rock, shale, and serpentine, which confirms the presence of ultramafic intrusives in the source area. The volcanic rocks are partially cryptocrystalline or undevitrified despite their antiquity. Phenocrysts of untwinned feldspar and quartz indicate silica-rich rocks, although some may be as mafic as andesite.

Fragments of "granitic" rocks (probably tonalite), "graphic granite," low-grade schist, and gneiss are not abundant, but they are widespread. A variety of limestone fragments is found, particularly in coarse-grained graywackes of type 3, but it is uncertain whether or not they were lithified prior to transport and deposition in the flysch trough. Partially recrystallized chert is present in most samples and the coarser-grained rocks contain siltstone and sandstone, including graywacke and metagraywacke. The presence of any sandstone fragments at all is significant because they are unlikely to survive weathering and erosion as sandstone fragments in sediments of comparable grain size; complete breakdown to monomineralic grains is more probable. Flattened argillite fragments can be assumed to have been soft sediment eroded locally from the bottom, but unconsolidated sand lacks the cohesion between grains necessary to maintain it as fragments. Fewer than ten sandstone clasts were found among the thousands of megascopic argillite and calcisiltite clasts, although unconsolidated graywacke was probably as available for erosion as was calcisiltite. The sandstone clasts which do not show soft-sediment deformation may have been derived from partially lithified sandstones outside the depositional area.

The currents which deposited the nonpelagic rocks of members $\alpha 1$ through $\gamma 2$ flowed from the east and southeast (Fig. 5). The change in direction from southeast in the lower members to east in the higher ones (Table 5) reflects the fact that successively higher members were studied further west along the salient trough in which currents were longitudinal, rather than indicating a shift of source area. All of the sediment did not necessarily originate from the southeastern end of the trough, wherever that may have been; the currents may have entered the trough from the sides and then turned and followed its longitudinal slope ("paleogeographic view" of Dzulynski and others, 1959). No member can be traced the length of the trough to show that current directions are constant throughout. It is possible that each member extended little beyond its present outcrop area and that the members represent imbricated shingles built out from a previously filled end of the trough or from a series of lateral deltas and submarine canyons.

The uppermost two members with current directions dominantly from the west require a major shift in the sediment supply line. The interlayering of beds deposited by currents flowing from the east and from the west at the base of $\gamma 3$ member shows that at least for a short period of time the trough was being filled from both ends. The alternation of currents from opposite ends of the trough should indicate the deepest portion of the trough since it is unlikely that the currents flowed for appreciable distances uphill.

The source area of the graywackes supplied considerable chromite-rich serpentine, felsic and other volcanic rocks, feldspathic plutonic rocks free of K-feldspar, chert, siltstone, and quartz-rich sedimentary rocks, including graywackes. This typical eugeosynclinal assemblage of graywacke, siltstone, chert, volcanic rocks, and serpentine raises the spectre of cannibalism. It appears that the Cloridorme Formation may have fattened its sedimentary pile at the expense of kindred flysch deposits elevated as tectonic lands within the eugeosyncline.

The most obvious source of these fragments is the Cap-des-Rosiers Formation, now exposed south and east of the Cloridorme belt and possibly extending northeastward to the flysch trough of western Newfoundland. These rocks may have been exposed to rapid erosion along the southern margin of the Cloridorme trough, particularly if the Shickshock trend of faults and serpentine intrusives extends east under the cover of Silurian and Devonian rocks. A positive Bouguer gravity anomaly associated with the exposed Shickshock rocks (Tanner and Uffen, 1960) increases in magnitude but becomes broader to the east (Fig. 1), probably due to greater depth of the excess mass, so that it cannot be followed as a distinct axis across northeastern Gaspé Peninsula where few gravity stations were observed. A broad gravity high crosses the coast between Cloridorme and Rivière-au-Renard, but equally large positive anomalies are found just north of the Dartmouth River at the west end of Gaspé Bay, apparently forming a trend which would lie a few kilometers north of Mt. Serpentine. An aeromagnetic survey map which will form an essential complement to the gravity study is in preparation. At present it appears that either or both of the areas of large positive anomalies in northeastern Gaspé could reflect buried ultramafic bodies and that the southern one, at least, could have provided the chromite and serpentine found in the Cloridorme Formation. Mt. Serpentine may be an exposed portion of the source if the intrusive is indeed as old as Ordivician (Alcock, 1924b; Neale and others, 1961, p. 2, 5) rather than Middle Devonian (Jones, 1935, p. 34). The presence of chromite and serpentine derived from the southeast in the Cloridorme Formation supports intrusion and exposure before the Middle Ordovician.

The properties of the marginal chromites of the Mt. Albert serpentine body in the Shickshock Mountains (MacGreggor and Smith, 1963, p. 409) are similar to those in the Cloridorme Formation and the date of intrusion, 495 m.y., is older than Middle Ordovician. However, when the reversal in current direction occurs in the $\gamma 3$ member which makes it easy to visualize the Shickshocks as the source area, the serpentine and chromite disappear. This suggests that the presently exposed serpentines of the Shickshock axis were not yet unroofed in the Middle Ordovician. The more quartz-rich metasedimentary rocks and the volcanic rocks of the Shickshock Group (Ollerenshaw, 1961; McGerrigle, 1954b) may have contributed.

Two dissonant notes in the general eugosynclinal nature of lithic fragments in the Cloridorme Formation rocks are the "granitic" rocks and the felsic nature of the volcanic rocks. The dark metavolcanic rocks of the Shickshock Group and the Cap-des-Rosiers Formation have been casually described as "basic volcanics" (Jones, 1934b, p. 46; Ollerenshaw, 1961, p. 4), basalts (McGerrigle, 1954b), and andesites (Alcock, 1924b, p. 138), although "a little quartz" was noted by Alcock (1924a, p. 130). Extensive quartz-rich metavolcanics have been mapped in the Ordovician of northern New Brunswick (Smith, 1967) and it may develop that they exist much closer at hand. Granitic rocks as old as the Cloridorme Formation are not known from northern Gaspé, but McGerrigle (1954b, p. 20) has described a conglomerate several hundred feet thick composed of "albite granite" clasts in the central Shickshock Mountains. The source of this unusual rock with 40 percent albite and 45 percent quartz is unknown, but its silicic composition and low K-feldspar content make either the conglomerate or its original source area an ideal donor of the "granite" and perhaps of quartz and feldspar in the Cloridorme Formation.

The source area of the calcisiltite and calcareous wacke supplied abundant carbonate mud (p. 18 *ff.*), but considerable quartz, feldspar, chromite, serpentine, and other terrigenous grains are also found in these rocks. The mineralogical components of graywackes and calcareous wackes are nearly identical, although in very different proportions, yet little intergradation between them is found. No distinction can be made between the carbonate and terrigenous rocks on the basis of current directions, which would help locate individual source areas, for they are statistically identical. The rather abrupt changes from nearly pure calcisiltite units to predominantly graywacke units must be explained while taking into account interlayering of some calcisiltites in most graywacke units.

The proportion of calcisiltite relative to pelagic argillite is not statistically different in those members which contain graywacke and in those which do not. This suggests that calcisiltite was supplied at a fairly constant rate averaged over the period of time represented by a member, if it is initially assumed that the rate of supply of pelagic argillite was essentially constant. This test is not particularly sensitive because the scatter of proportions is too high, possibly meaning that the initial assumption of constancy of rate of argillite supply is not valid. Furthermore, the same result is obtained whether or not the amount of argillite in each member is corrected to eliminate turbidite argillite, although the confidence level is somewhat higher if the correction is made. The relative proportions of inferred turbidite argillite to graywacke and to calcisiltite plus graywacke determined in the $^{\beta}7$ member (Enos, 1969) was used to make correction in other members.

Carbonate rocks older than the Cloridorme Formation are now exposed to the north (Mingan Islands, Twenhofel, 1938), south (Cambrian limestones, Alcock, 1935) and east (Newfoundland, Schuchert and Dunbar, 1934), but no contemporaneous carbonates are exposed, although they could exist in Jacques Cartier Passage between Anticosti and the Mingan Islands (Fig. 2). Despite this technicality, origin of the Cloridorme carbonate from a contemporaneous carbonate shelf rather than from erosion of pre-existing carbonate rocks is favored because the uniform microcrystalline texture of the carbonate in the many small pockets which have escaped recrystallization suggests calcareous-mud rather than lithoclastic origin. A shallow shelf partially isolated from terrigenous influx could supply carbonate mud into deeper areas through bordering canyons by periodic slumping or storm agitation.

The mineralogical similarities between the graywackes and calcisiltites, particularly the omnipresence of chromite and serpentine, suggest that the postulated carbonate shelf may have lain adjacent to and only partially isolated from the terrigenous source. Coarse terrigenous material could be carried through the carbonate shelf into the flysch trough in submarine canyons which headed near land, perhaps at a river mouth; canyons heading at the edge of the shelf would supply carbonate to the trough. The island of Cuba with flanking carbonate shelves, rugged interior uplands, abundant serpentines, and nearby deep trenches may be a modern counterpart. The situation may also have been one of lateral separation between terrigenous and carbonate sources. The Persian Gulf with its isolated lateral carbonate shelves and the Tigris-Euphrates drainage delivering abundant material from the end is a modern example. Both of these arrangements for waste disposal could mix the terrigenous and carbonate end members either on the shelf by encroachment of terrigenous material, in the canyons, or on the floor of the flysch trough by mixing of currents and by current erosion of the bottom. The uniform orientation of current directions in the Cloridorme trough suggests that if the two types of sediment were supplied by areally separated canyons, the canyon mouths must have been distant enough from the final resting place of the sediment to allow the slope of the trough to channel all currents into the same longitudinal axis.

To explain the periodic rather abrupt changes from pure calcisiltite deposition to predominantly graywacke deposition, the source model involving areal separation of the canyons supplying carbonate and those supplying graywacke is preferred. The canyons supplying carbonate must have been closer to the rocks now preserved so that at times carbonate unadulterated by graywacke could be supplied. The supply of carbonate was presumably less variable than that of terrigenous material, or at any rate subject to different controls such as water temperature and surface currents. The major control of the supply of terrigenous material would be the rates of uplift and erosion in the source area. During times of abundant supply, the terrigenous material would build out as a wedge or submarine fan into areas where carbonate sedimentation previously dominated. Each successive current would flow down the slope provided by the wedge of sediment deposited by its predecessors and deposit its load as a new shingle at the edge of the pile. The intertonguing of graywacke and calcisiltite at the base of $\beta 7$ member (Fig. 6) appears to be caused by wedges of graywacke which thin westward. With a decrease in the supply of terrigenous material the terrigenous wedge would find its supply lines overextended and all of the terrigenous material would be consumed in maintaining the slope nearer the source area. Calcareous material would not be subject to the same fluctuations and could regain ascendancy in the further reaches of the trough with the decline in abundance of terrigenous material. The mechanism for interlayering of isolated calcisiltite beds in graywacke units is built into this explanation. The graywacke units represent an overwhelming of the carbonate supply with terrigenous material, not a cessation in carbonate supply. Carbonate would continue to be supplied at about the same rate, as it appears that it was (p. 50), but its relative importance would be reduced to scattered interbeds.

Flysch Facies

The Cloridorme Formation belongs to the flysch lithofacies as defined here or by most recent authors (p. 3), but it is also flysch by the other common concepts of the term.

Flysch as a geologic term was borrowed by Studer (1827, p. 39) from a German Swiss dialect word for distinctive rocks consisting of alternating thin, uniform beds of sandstone, dark shale, and limestone in the northwestern Alps. Early use of the word in the Alps tended to restrict it to the characteristic rocks of early Tertiary age, although Studer had already applied it to Cretaceous and Jurassic rocks as well (Tercier, 1947), but it is now used solely as a facies term. Diverse usages as to types of facies have evolved, however (Eardley and White, 1947). Following Bertrand's (1897) generalization that Alpine flysch was deposited in the last stages of the geosyncline prior to an orogenic climax, flysch has been used as a tectofacies. Tercier (1947), Sujkowski (1957), and Dzulynski (1963) stress the lithofacies aspect of flysch. It is now apparent that turbidity currents are involved in the deposition of many rocks of this lithologic character and structural setting (Kuenen and Migliorini, 1950; Kuenen, 1958; Dzulynski, 1963), and this type of deposition is implicit in many geologists' use of the term. Sanders and Carozzi (1957) stress deep-water origin of flysch and would use "flysch" to replace "eugosynclinal" as applied to clastic rocks. However, all eugosynclinal clastic rocks are not flysch as these terms are most commonly used, and a reliable determination of depth of deposition is rarely possible. Some flysch is characterized by pelagic fossils, identifiable as cold- or deep-water forms in some younger rocks, and reworked shallow-water fossils in the coarser clastics (Natland and Kuenen, 1951). Seilacher (1954, 1959) has distinguished a suite of feeding marks as characteristic of flysch deposits. Thus the word may also be used for a biofacies.

The lithofacies aspect of the Cloridorme flysch has been documented. The pelagic and transported faunal relations are those commonly found in flysch and the rare feeding marks may be those of the flysch biofacies of Seilacher (1954, 1959). Petrologic evidence indicates that uplift in the Lower Ordovicain flysch trough (Cap-des-Rosiers Formation, Shickshock Group?) had begun during the deposition of the Cloridorme Formation. The Cloridorme Formation was itself involved in orogenesis before the Middle Silurian (Cumming, 1959; Crickmay, 1932) and presumably before the Late Ordovician by comparison with the less-deformed Pabos-White Head Limestone of central Gaspé and by analogy to the age of the Taconic orogeny elsewhere in the Hudson Valley-Gaspé flysch trough (Zen, 1961; Henderson, 1958; Sanders and others, 1961). The Cloridorme Formation was deposited in the last stages of geosynclinal development prior to an orogeny; it is therefore of the flysch tectofacies (Bertrand, 1897; Tercier, 1947) as well.

References Cited

Alcock, F. J., 1924a, Shickshock Mountains, Central Gaspé, Quebec: Canada Geol. Survey Summary Rept., pt. C., 127-133.

—— 1924b, Geology of Mount Serpentine, Gaspé, Quebec: Canada Geol. Survey Summary Rept., pt. C, p. 134-141.

—— 1935, Geology of the Chaleur Bay region: Canada Geol. Survey Mem. 183, 146 p.

—— 1941, Jacquet River and Tetagouche River map-areas, New Brunswick: Canada Geol. Survey Mem. 227, 43 p.

Allen, J. R. L., and Narayan, J., 1964, Cross-stratified units, some with silt bands, in the Folkestone Beds (Lower Greensand) of southeast England: Geol. en Mijnb., v. 43, p. 451-461.

Ayrton, W. G., 1961, Chandler-Port Daniel area: Quebec Dept. Mines Prelim. Rept. 447, 12 p.

Bailey, E. H., and Irwin, W. P., 1959, K-feldspar content of Jurassic and Cretaceous greywackes of northern Coast Ranges and Sacramento Valley, California: Am. Assoc. Petroleum Geologists Bull., v. 43, no. 12, p. 2797-2809.

Béland, Jacques, 1957, Ste.-Félicité-Grosses Roches Area: Quebec Dept. Mines Prelim. Rept. 339, 8 p.

Bell, H. S., 1942, Density currents as agents for transporting sediments: Jour. Geology, v. 50, p. 512-547.

Belt, E. S., 1963, Stratigraphy and sedimentology of the Mabou Group (Middle Carboniferous), Nova Scotia, Canada: Ph.D. dissert., New Haven, Connecticut, Yale Univ.

Berry, W. B. N., 1960, Graptolite faunas of the Marathon region, West Texas: Texas Univ. Pub. 6005, 179 p.

—— 1962a, On the Magog, Quebec, graptolites: Am. Jour. Sci., v. 260, p. 142-148.

—— 1962b, Stratigraphy, zonation, and age of Schaghticoke, Deepkill and Normanskill Shales, eastern New York: Geol. Soc. America Bull., v. 73, p. 695-718.

Bertrand, Marcel, 1897, Structure des Alpes francaises et récurrence de certain faciès sedimentaires: Congrès géol. intern., Compte Rendu, 16th sess., 1894, p. 161-177.

Bokman, John, 1955, Sandstone classification: relation to composition and texture: Jour. Sed. Petrology, v. 25, p. 201-206.

Bouma, A. H., 1961, Sedimentology of some flysch deposits; a graphic approach to facies interpretation: Ph.D. dissert., Univ. Utrecht, The Netherlands.

—— 1962, Sedimentology of some flysch deposits; a graphic approach to facies interpretation: Amsterdam, Elsevier, 168 p.

Cady, W. M., 1960, Stratigraphic and geotectonic relationships in northern Vermont and southern Quebec: Geol. Soc. America Bull., v. 71, p. 531-576.

Cooke, H. C., 1950, Geology of a southwestern part of the Eastern Townships of Quebec: Canada Geol. Survey Mem. 257, 142 p.

Cooper, G. A., 1956, Chazyan and related brachiopods: Smithsonian Misc. Coll., v. 127, p. 1-1245.

Crickmay, G. W., 1932, Evidence of Taconic orogeny in the Matapedia Valley, Quebec: Am. Jour. Sci., ser. 5, v. 24, p. 368-386.

Crook, K. A. W., 1960, Classification of arenites: Am Jour. Sci., v. 258, p. 419-428.

Crowell, J. C., 1955, Directional-current structures from the Prealpine Flysch, Switzerland: Geol. Soc. America Bull., v. 66, p. 1351-1384.

Cumming, L. M., 1959, Silurian and Lower Devonian formations in the eastern part of Gaspé Peninsula, Quebec: Canada Geol. Survey Mem. 304, 45 p.

Cummins, W. A., 1958, Sedimentary structures from the lower Keuper Sandstone: Liverpool-Manchester Geology Jour., v. 2, pt. 1, p. 37-43.

—— 1962, The greywacke problem: Liverpool-Manchester Geology Jour., v. 3, pt. 1, p. 51-72.

Currier, L. W., and Jahns, R. H., 1941, Ordovician stratigraphy of central Vermont: Geol. Soc. America Bull., v. 52, p. 1487-1512.

Dapples, E. C., Krumbein, W. C., and Sloss, L.L., 1953, Petrographic and lithologic attributes of sandstones: Jour. Geology, v. 61, p. 291-317.

Dunbar, C. O., and Rodgers, John, 1957, Principles of stratigraphy: New York, John Wiley and Sons, Inc., 356 p.

Dunham, R. J., 1962, Classification of carbonate rocks according to depositional texture, p. 108-121 *in* Ham, W. E., *Editor,* Classification of carbonate rocks—a symposium: Am. Assoc. Petroleum Geologists Mem. 1.

Dzulynski, St., 1963, Directional structures in flysch: Studia Geologica Polonica, v. XII, p. 81-136.

Dzulynski, St., and Sanders, J. E., 1962, Current marks on firm mud bottoms: Connecticut Acad. Arts and Sci., Trans. v. 42, p. 57-96.

Dzulynski, St., Ksiaskiewicz, M., and Kuenen, Ph. H., 1959, Turbidites in flysch of the Polish Carpathian Mountains: Geol. Soc. America Bull., v. 70, p. 1089-1118.

Dzulynski, St., Radomski, A., and Slaczka, A., 1957, Sandstone whirlballs in the silts of the Carpathian-Flysch: Ann. Soc. Geol. Pologne, v. 26 (1956), p. 107-126.

Eardley A. J., and White, M. G., 1947, Flysch and molasse: Geol Soc. America Bull., v. 58, p. 979-990

Ells, R. W., 1883, Report on the geological formations of the Gaspé Peninsula: Canada Geol. Survey Rept. Prog., 1880-81-82, pt. D, 32 p.

—— 1884, Report on exploration and surveys in the interior of the Gaspé Peninsula: Canada Geol. Survey Rept. Prog., 1882-83-84, pt. E, 34 p.

Emery, K. O., 1950, Contorted Pleistocene strata at Newport Beach, California: Jour. Sed. Petrology, v. 20, p. 111-115.

Enos, Paul, 1965, Discontinuity of turbidite beds, Gaspé Peninsula, Canada (Abstract): Geol. Soc. America Spec. Paper 87, p. 51.

—— 1969, Anatomy of a Flysch, Jour. Sed. Petrology, v. 39 (in press).

Ericson, D. B., Ewing, M., and Heezen, B. C., 1952, Turbidity currents and sediments in North Atlantic: Am. Assoc. Petroleum Geologists Bull., v. 36, p. 489-511.

Folk, R. L., 1954, The distinction between grain size and mineral composition in sedimentary rock nomenclature: Jour. Geology, v. 62, p. 344-359.

—— 1959, Practical petrographic classification of limestones: Am. Assoc. Petroleum Geologists Bull., v. 43, p. 1-38.

Gilbert, C. M., 1954, Sedimentary rocks, p. 251-384 *in* Williams, Howel, Turner, F. J., and Gilbert, C. M., Petrography: San Francisco, W. H. Freeman and Co., 406 p.

Gilbert, G. K., 1914, The transportation of debris by running water: U.S. Geol. Survey Prof. Paper 86, 263 p.

Hamblin, W. K., 1962, X-ray radiography in the study of structures in homogeneous sediments: Jour. Sed. Petrology, v. 32, no. 3, p. 201-210.

Henderson, W. R. S., 1958, "Blountian" allochthone in Appalachians of Quebec: Alberta Soc. Petroleum Geologists Jour., v. 6, p. 120-128.

Hubert, J. F., 1964, Textural evidence for deposition of many western North Atlantic dep-sea sands by ocean-bottom currents rather than turbidity currents: Jour. Geology, v. 72, no. 6, p. 757-785.

Hutchinson, R. D., 1952, Middle Cambrian of the Atlantic realm in eastern Gaspé: Am. Jour. Sci., v. 250, p. 275-280.

Hutton, C. O., 1950, Studies of heavy detrital minerals: Geol. Soc. America Bull., v. 61, no. 7, p. 635-710.

Jones, I. W., 1934a, Marsoui map-area, Gaspé Peninsula: Quebec Bur. Mines Ann. Rept. for 1933, pt. D, p. 3-40.

—— 1934b, Summary report on north-central Gaspé: Quebec Bur. Mines Ann. Rept. for 1933, pt. D, p. 41-54.

—— 1935, Dartmouth River map-area, Gaspé Peninsula: Quebec Bur. Mines Ann. Rept. for 1934, pt. D, p. 3-44.

Jopling, A. V., 1964; Laboratory study of sorting processes related to flow separation: Jour. Geophys. Research, v. 69, no. 16, p. 3403-3418.

Kaye, A., and Power, W. R., Jr., 1954, A flow cast of very recent date from northeastern Washington: Am. Jour. Sci., v. 252, p. 309-310.

Kindle, E. M., 1938, The correlation of certain Devonian fauna of eastern and western Gaspé: Am. Assoc. Paleontologists Bull. v. 24, no. 82, 50 p.

Klein, G. de V., 1963, Analysis and review of sandstone classifications in the North American geological literature, 1940-1960: Geol. Soc. America Bull., v. 74, no. 5, p. 555-576.

Krumbein, W. C., 1941, Measurement of geological significance of shape and roundness of sedimentary particles: Jour. Sed. Petrology, v. 11, no. 2, p. 64-72.

Krumbein, W. C., and Garrels, R. M., 1952, Origin and classification of chemical sediments in terms of *p*H and oxidation-reduction potentials: Jour. Geology, v. 60, no. 1, p. 1-33.

Krynine, P. D., 1948, The megascopic study and field classification of sedimentary rocks: Jour. Geology, v. 56, p. 130-165.

Ksiazkiewicz, M., 1952, Graded and laminated bedding in the Carpathian flysch: Polskiego Towarzystwa Geologicznego, Rocznik, Krakow, v. 22, p. 399-499.

Kuenen, Ph. H., 1953, Significant features of graded bedding: Am. Assoc. Petroleum Geologists Bull., v. 37, p. 1044-1066.

—— 1957, Sole marking of graded greywacke beds: Jour. Geology, v. 65, p. 231-238.

—— 1958, Problems concerning source and transportation of flysch sediments: Geol. en Mijnb., nw. ser., v. 20, p. 329-339.

—— 1959, Experimental abrasion of pebbles, 3. Fluviatile action on sand: Am. Jour. Sci., v. 257, p. 172-190.

Kuenen, Ph. H., and Migliorini, C. I., 1950, Turbidity currents as a cause of graded beddings: Jour. Geology, v. 58, p. 91-127.

Kuenen, Ph. H., and Sanders, J. E., 1956, Sedimentation phenomena in Kulm and Flözleeres graywackes, Sauerland and Oberharz, Germany: Am. Jour. Sci., v. 254, p. 649-671.

Logan, W. E., 1846, Sections on Chaleur Bay and the coast of Gaspé: Canada Geol. Survey Rept. Prog., 1844, p. 78-110.

—— 1863, Geology of Canada: Canada Geol. Survey Rept. Prog. from commencement to 1863, 983 p.

Love, L. G., 1962a, Biogenic primary sulfide of the Permian Kupferschiefer and Marl Slate: Econ. Geology, v. 57, p. 350-366.

—— 1962b, Pyrite spheres in sediments, p. 121-143, *in* Jensen, M. L., *Editor,* Biogeochemistry of sulfur istotopes: New Haven, Connecticut, Yale Univ., N.S.F. Symposium.

MacGreggor, I. D., and Smith, C. H., 1963, The use of chrome spinels in petrographic studies of ultramafic intrusions: Canada Mineralogist, v. 7, pt. 3, p. 403-412.

McBride, E. F., 1962, Flysch and associated beds of the Martinsburg Formation (Ordovician), Central Appalachians: Jour. Sed. Petrology, v. 32, no. 1, p. 39-91.

McGerrigle, H. W., 1950, The Geology of Eastern Gaspé: Quebec Dept. Mines Geol. Rept. 35, 174 p.

McGerrigle, H. W., 1953, Geology of Gaspé Peninsula: Quebec Dept. Mines, Map, No. 1000.

—— 1954a, An outline of the geology of Gaspé Peninsula: Canadian Mining Jour., v. 75, no. 8, p. 57-63.

—— 1954b, The Tourelle and Courcelette areas: Quebec Dept. Mines Geol. Rept. 62, 63 p.

—— 1959, Madeleine River area: Quebec Dept. Mines Geol. Rept. 77, 50 p.

McKee, E. D., and Weir, G. W., 1953, Terminology for stratification and cross-stratification in sedimentary rocks: Geol. Soc. America Bull., v. 64, p. 381-390.

Menard, H. W., 1950, Current-ripple profiles and their development: Jour. Geology, v. 58, p. 152-153.

Middleton, Gerard V., 1962, Size and sphericity of quartz grains in two turbidite formations: Jour. Sed. Petrology, v. 32, no. 4, p. 725-742.

Moore, R. C., Frye, J. C., Jewett, J. M., Lee, Wallace, O'Connor, H. G., 1951, The Kansas rock column: Kansas Geol. Survey Bull. 89, 132 p.

Natland, M. L., and Kuenen, Ph. H., 1951, Sedimentary history of the Ventura Basin, Calif., and the action of turbidity currents, p. 76-107 *in* Turbidity currents: Soc. Econ. Paleontologists and Mineralogists Spec. Pub. no. 2.

Neale, E. R. W., Béland, J., Potter, R. R., and Poole, W. H., 1961, A preliminary tectonic map of the Canadian Appalachians based on age of folding: Canadian Mining and Metallurgy Bull., 8 p.

Ollerenshaw, N C., 1961, Cuog-Langis Area: Quebec Dept. Mines Prelim. Rept. no. 465, 14 p.

Osborne, F. F., 1956, Geology near Quebec City: Le Naturaliste Canadien, v. 83, no. 8-9, p. 157-223.

Packham, G. H., 1954, Sedimentary structures as an important factor in the classification of sandstones: Am. Jour. Sci., v. 252, p. 466-476.

Passega, R., 1957, Texture as charcteristics of clastic deposition: Am. Assoc. Petroleum Geologists Bull., v. 41, p. 1952-1984.

Pettijohn, F. J., 1954, Classification of sandstones: Jour. Geology, v. 62, p. 360-365.

Potter, P. E., and Glass, H. D., 1958, Petrology and sedimentation of Pennsylvanian sediments in southern Illinois: A profile: Illinois Geol. Survey Rept. Inv. 204, 60 p.

Potter, P. E., and Pettijohn, F. J., 1963, Paleocurrents and basin analysis: New York, Academic Press, 296 p.

Potter, P. E., and Pryor, W. A., 1961, Dispersal centers of Paleozoic and later clastics of the upper Mississippi Valley and adjacent areas: Geol. Soc. America Bull., v. 72, no. 8, p. 1195-1250.

Ramsay, J. G., 1961, The effects of folding upon the orientation of sedimentation structures: Jour. Geology, v. 69, p. 84-100.

Rich, J. L., 1950, Flow markings, groovings, and intrastratal crumplings as criteria for recognition of slope deposits, with illustrations from Silurian rocks of Wales: Am. Assoc. Petroleum Geologists Bull., v. 34, p. 717-741.

Rodgers, John, 1957, Geologic map of east Tennessee with explanatory text: Tennessee Div. Geol. Bull. 58, pt. 2, 168 p.

Ruedemann, R., 1901, Hudson River beds near Albany and their taxonomic equivalents: New York State Mus. and Sci. Service Bull. 42, p. 489-568.

Sanders, J. E., 1959, Symbols for graphic representation of directional sedimentary structures (Abstract): Geol. Soc. America Bull., v. 70, no. 12, p. 1667-1668.

—— 1960, Origin of convolute lamination: Geol. Mag., v. 97, p. 409-421.

—— 1963, Concepts of fluid mechanics provided by primary sedimentary structures: Jour. Sed. Petrology, v. 33, no. 1, p. 173-179.

—— 1965, Primary sedimentary structures formed by turbidity currents and related resedimentation mechanisms, p. 192-219 *in* Middleton, G. V., *Editor,* Primary sedimentary structures and their hydrodynamic interpretation: Soc. Econ. Paleontologists and Mineralogists Spec. Pub. 12.

Sanders, J. E., and Carozzi, A. V., 1957, Flysch and molasse (Abstract): Geol. Soc. America Bull., v. 68, p. 1790-1791.

Sanders, J. E., Platt, Lucian, and Powers, R. W., 1961, Bald Mountain Limestone, New York: New facts and interpretations relative to Taconic geology: Geol. Soc. America Bull., v. 72, p. 485-488.

Schuchert, Charles, and Dunbar, C. O., 1934, Stratigraphy of western Newfoundland: Geol. Soc. America Mem. 1, 122 p.

Seilacher, Adolf, 1954, Die geologische Bedeutung fossiler Lebensspuren: Deutsch. Geol. Gesell. Zeitschr., bd. 105 (1953), p. 214-227.

—— 1959, Zur ökologischen Characteristik von Flysch und Molasse: Eclogae Geol. Helvetiae, v. 51 (1958), p. 1062-1078.

Shepard, F. P., 1951, Transportation of sand into deep water, p. 53-65 *in* Turbidity currents: Soc. Econ. Paleontologists and Mineralogists Spec. Pub. no. 2.

Smith, C. H., 1957, Bathhurst-Newcastle area, New Brunswick: Canada Geol. Survey Map 1 - 1957.

—— 1958, Bay of Islands igneous complex, western Newfoundland: Canada Geol. Survey Mem. 290, 132 p.

Sorby, H. C., 1859, On the structures produced by the currents present during the deposition of stratified rocks: The Geologist, v. 2, p. 137-147.

Spurr, J. E., 1894, False bedding in stratified drift deposits: Am. Geologist, v. 13, p. 43-47.

Studer, B., 1827, Geognostische Bemerkungen über einige Theile der nordlichen Alpenkette: Zeitschr. Mineralogie, Jahrg. 1827, no. 1, p. 1-46.

Sujkowski, Zb. L., 1957, Flysch sedimentation: Geol. Soc. America Bull., v. 68, p. 543-554.

Tanner, J. G., and Uffen, R. J., 1960, Gravity anomalies in Gaspé Peninsula, Quebec: Dominion Observatory Ottawa Pubs. v. 21, no. 5, p. 220-237.

ten Haaf, E., 1956, Significance of convolute lamination: Geol. en Mijnb., nw. ser., v. 18, p. 188-194.

—— 1959, Graded beds in the northern Appennines: Ph.D. dissert., Rijksuniversiteit Groningen, Groningen, The Netherlands, 102 p.

Tercier, J., 1947, Le flysch dans le sedimentation alpine: Eclogae Geol. Helvetiae, v. 40, no. 2, p. 163-198.

Twenhofel, W. H., 1928, Geology of Anticosti Island: Canada Geol. Survey Mem. 154, 481 p.

—— 1938, Geology and paleontology of the Mingan Islands: Quebec: Geol. Soc. America Spec. Paper 11, 132 p.

Vallentyne, J. R., 1962, A chemical study of pyrite spherules isolated from sediments of Little Round Lake, Ontario, p. 144-152 *in* Jensen, M. L., *Editor,* Biogeochemistry of sulfur isotopes: New Haven, Connecticut, Yale Univ., N.S.F. Symposium.

van Andel, T. H., 1958, Origin and classification of Cretaceous, Paleocene, and Eocene sandstones of western Venezuela: Am. Assoc. Petroleum Geologists Bull., v. 42, p. 734-763.

Walker, R. G., 1963, Distinctive types of ripple-drift cross-lamination: Sedimentology, v. 2, no. 3, p. 173-188.

Wood, Alan, and Smith, A. J., 1958, Two undescribed structures in a greywacke series: Jour. Sed. Petrology, v. 28, p. 97-101.

Zen, E-an, 1961, Stratigraphy and structure at the north end of the Taconic Range in west-central Vermont: Geol. Soc. America Bull., v. 72, no. 2, p. 293-338.

MANUSCRIPT RECEIVED BY THE SOCIETY DECEMBER 12, 1966

Author Index

Subject Index